CI/SfB Construction inde

Cover:

The proper organisation of the information which members of the building team use and produce is a means to architecture, not an end in itself.

Architects Faulkner-Brown Hendy Watkinson Stonor of Newcastle upon Tyne use CI/SfB in conjunction with the RIBA Plan of Work for the systematic arrangement and co-ordination of information.

On their Bletchley Leisure Centre project, shown on the cover, CI/SfB was used for the organisation of production drawings and for specifications.

Photograph by J. Reid, published in *Building,* 28 February 1975.

The authors:

Alan Ray-Jones BA Arch (UCL) RIBA has worked in public and private practice, and for the RIBA. He has been responsible for the SfB Agency since 1965, and represents the Agency on the CIB Committee concerned with the international development and promotion of SfB.

David Clegg ALA has worked in the construction industry since 1967, first in West Sussex County Architect's Department, and from 1969 with RIBA Services Ltd. He is involved in all aspects of the work of the SfB Agency.

Construction indexing manual

Alan Ray-Jones RIBA and David Clegg ALA
SfB Agency UK

RIBA Publications Limited

First edition published 1961 (SfB/UDC Building Filing Manual)
Second revised edition published 1968 (Construction Indexing Manual)
Third revised edition published 1976 (CI/SfB Construction Indexing Manual)

© RIBA 1976   Reprinted 1978, 1982

ISBN: 0 900630 51 5

Design and production by Michael Stribbling
and Gordon Burrows

Typesetting by Jolly & Barber Ltd, Rugby
and Vantage Photosetting Co Ltd, Southampton

Printed by Balding & Mansell,
Wisbech, Cambs

Published by RIBA Publications Limited
Finsbury Mission, Moreland Street, London EC1V 8VB, England

This publication is copyright under the Berne Convention and the International Copyright Convention. All rights reserved. Apart from any fair dealing under the UK Copyright Act 1956, part 1, section 7, whereby a single copy of an article may be supplied, under certain conditions, for the purposes of research or private study, by a library of a class prescribed by the UK Board of Trade Regulations (Statutory Instruments 1957, No. 868), no part of this publication may be reproduced, stored in a retrieval system or transmitted in any form or by any means without the prior permission of the copyright owners. Multiple copying of the contents of the publication without permission is always illegal.

Contents

Page 7 Foreword
8 Acknowledgements
9 Using CI/SfB

Tables

16 Introduction to the tables
19 Table 0 Physical environment
35 Table 1 Elements
57 Table 2 Constructions, forms
63 Table 3 Materials
69 Table 4 Activities, requirements
89 Index to the tables

Applications

123 1 Project information
123 1.1 Introduction
126 1.2 Design
132 1.3 Production ('working') drawings
142 1.4 Specification and quantities
148 1.5 Other project information applications

154 2 General information
154 2.1 Introduction
160 2.2 Published trade and technical literature
162 2.3 Office libraries

Appendices

167 1 Common element parts
171 2 Classification and filing
187 3 Indexing
195 4 Theoretical basis of CI/SfB
200 5 History of SfB
205 6 Correlation with other systems

To have a perfect system is impossible; to have a system is indispensable.
G. K. Chesterton

CI/SfB Construction indexing manual

The authoritative United Kingdom version of SfB (third edition) for use in project information and related general information with a guide to its use for project data coordination, office libraries and other applications.

CI/SfB is in use for the coordination of information by offices with one or two staff up to one to two thousand staff, for all sizes of projects, for new work and alterations. The initials CI stand for 'Construction Index'. SfB stands for 'Samarbetskommittén för Byggnadsfrågor', the name of the classification system authorized by the International Council for Building Research Studies and Documentation (CIB) for the structuring and filing of construction industry information.

Foreword

The foreword I wrote for the last edition of this manual referred to the potential use of SfB. This one marks a considerable achievement in which the RIBA, in sponsoring SfB in the United Kingdom, plays a leading role.

SfB is already used for the arrangement of most office libraries and for product information in the UK. Research work carried out by the Building Research Establishment has shown that the system can provide a satisfactory means of structuring sets of working drawings. It is also used for the arrangement of the National Building Specification, and is aligned closely with the Standard Form of Cost Analysis. A new development which will be watched with special interest by the member institutes of CIB is the proposal for the re-arrangement of the Scottish Building Regulations, which reflects the influence of SfB and could provide a means of standardizing the presentation of legislative requirements related to building.

From the point of view of CIB, I welcome the fact that almost all the codes and headings given in Tables 1, 2 and 3 in this edition of the Construction indexing manual are identical with their equivalents in CIB Report 22. There is no doubt that this virtual elimination of basic code differences will encourage the further international use of SfB.

I am happy to say that the work which the RIBA is doing is being actively taken up by organisations in other countries including those in France, Germany and other European nations as well as in Asia, Africa, South America and elsewhere. This manual should now be adopted in preference to all previous editions. I realise that many established users will regret the need for a new edition, but I am sure that they will acknowledge, as I do, that the changes made now are much less sweeping than those made in the 1968 edition. Change is always painful for those who have already adopted a system, but inevitable if it is to continue to flourish and adapt successfully to new circumstances. I wish the manual every success.

Ingvar Karlén
Chairman, CIB Commission W52

Stockholm 1975

Acknowledgements

This book is the third edition of the United Kingdom SfB Manual, previous editions being the 'SfB/UDC Building Filing Manual' (1961), and the 'CI/SfB Construction indexing manual' (1968).

SfB evolves in the light of experience, so our thanks are due to all those associated with the two previous editions, including Dargan Bullivant RIBA who guided the development of SfB in the United Kingdom until 1968 and to Lars Magnus Giertz, who founded SfB in 1949 and has taken a leading part in discussions in CIB's SfB Development Group.

In developing CI/SfB since 1968 and preparing this manual, we have benefited from advice given to us by the UK SfB Committee 1972 and by its members individually. The SfB Committee acted as an informal advisory group of those with related interests and responsibilities able to bring different points of view to bear on the work of revision.

We have similarly, through contact with CIB Commission W31, had the benefit of the views of those concerned principally with the listing of properties for buildings and their parts. Many of the recommendations for the application of CI/SfB to working drawings are based on BRE research reported in Current Paper 18/73 and BRE Digest 172.

We have made contact with specialist organisations and with a large number of users of the 1968 edition, both in this country and overseas. All these organisations and individuals have given generously of their experience and have gone out of their way to give detailed comment on development schedules and drafts.

SfB Committee 1972

A. M. Lewis, Managing Director RIBAS and NBS (Chairman)
R. A. Allott RIBA, National Building Specification
S. Bell RIBA, Building Industry Code
B. Drake FRICS, Building Cost Information Service
B. C. Edgill ARICS, NCC Working Party on Data Coordination
M. Jeffery RIBA, Greater London Council
J. Mills FLA, Classificationist, The Polytechnic of North London
A. Ray-Jones RIBA, RIBA Services
M. Roberts ALA, Classificationist, Construction Industry Thesaurus
D. Robertson ARICS, Building Cost Information Service
R. Swanston ARICS, Building Industry Code
D. Clegg ALA, SfB Agency UK (Secretary)

The illustrations in the schedules are by Wilf McCann and David Martin. Many illustrations in the Applications section are based on sketches or original ideas by Wilf McCann, and drawings in Using CI/SfB 1.2, 1.3 and Applications 1.1, 2.1 are by David Martin.

Finally, we would like to thank Dorcas Finch and Yasmine Roberts for typing successive editions of the manual in draft.

Alan Ray-Jones RIBA, SfB Agency UK
David Clegg ALA, SfB Agency UK

Using CI/SfB

1.1 If you already know CI/SfB . . .

The main influences on CI/SfB development since publication of the last edition of the Construction indexing manual in 1968 and the Project manual in 1971 have been the experience of users; international exchange of experience and discussion within CIB; the development of the National Building Specification (NBS); the BRE research on sets of working drawings; the studies sponsored by the National Consultative Council (NCC) Working Party on Data Co-ordination including the development of the Construction Industry Thesaurus (CIT); and the revision of the CIB Master List of Properties. Most of the changes made in this edition, unlike the last, are to the detail rather than the structure of CI/SfB; the index has also been improved. The Construction Industry Thesaurus has been used throughout the development work and has considerably assisted it.

The most important changes are listed below:

Table 0 **Physical environment (formerly 'Built Environment')**
The duplication of subject headings between 'Civil engineering works' and 'Transport, industrial buildings' has been removed. The new version of the table provides an improved means of classifying spaces in and around buildings, for the organisation of briefing information as well as library information.

Table 1 **Elements**
Element definitions have been brought more closely into line with the Standard Form of Cost Analysis (SFCA). A list of functional parts common to two or more elements has been developed and is included at Appendix 1 for specialist users only. It provides more detail than is normally required for general use in non-specialist offices.

Table 2 **Constructions, forms (formerly 'Construction Form')**
The use of CI/SfB for specification has led to a number of changes in Table 2. Constructions (eg, brickwork) and product forms (eg, bricks) are shown separately in this edition.

Table 3 **Materials**
This table may now be used on its own, not necessarily in combination with Table 2.

Table 4 **Activities and requirements**
The last edition of the Construction indexing manual included a list of properties based on the 1964 CIB Master List of Properties for Building Materials and Products. The CIB Master List was considerably improved in 1972 and several major subjects changed position. In particular, 'fire' was brought forward in the list and 'heat, light and sound' were brought together. The new edition of the master list was included as part of British Standard BS 4940 'The presentation of technical information about products and services in the construction industry' in 1973.

In view of the changes to the Master List and their acceptance in the UK, the SfB Agency considered it essential to bring CI/SfB into line, although it is realised that changes to Table 4 must inconvenience, in particular, librarians with detailed indexes. Wherever possible, major classes in Table 4 have been left unchanged, and the opportunity has been taken to omit the code symbol (l), to avoid confusion with (1).

This edition of CI/SfB will remain valid at least until 1984, although corrections and supplementary advice will be given in the RIBA 'Product Data' information system. In the period up to 1984, the SfB Agency will improve the schedules relating to management at (A), (B), (C) and (D). Co-ordination with the Construction Industry Thesaurus will be further developed.

All who are interested in the further development of the system or its application are invited to write to:
SfB Agency UK, 66 Portland Place London W1N 4AD
which is responsible for the administration, development and promotion of SfB in the United Kingdom.

1.2 If CI/SfB is new to you, or if you want to get up to date . . .

CI/SfB is a standard, common framework or list of headings, used for setting up office libraries, collecting design information, preparing reports, cost plans, drawings, specifications, bills of quantities and other types of information for building. The system has a long history of development (see Appendix 5) and is established internationally through the International Council for Building Research Studies and Documentation (CIB). Its strength lies in its flexibility. It can be used by small and large architectural firms; it can be used by quantity surveyors, engineers and contractors and it provides a basis for building industry coordination and communication.

Organisations associated with the building industry vary considerably in their size and working methods. The offices of architects, consultants, quantity surveyors, and builders all range from the minute to the enormous and their methods of working reflect this diversity of size and discipline. Although many architectural offices are relatively small, with slender resources, their functions and responsibilities are essentially the same as those of large practices. Jobs may be smaller and fewer and overheads less, but some system of organising information is essential. Every practitioner has a collection of incoming technical information and has to organise the project information he produces, to a reasonable standard at a reasonable cost.

At the other extreme, large practices also need administrative, technical, and information services. Their problems are different in detail; they may have a librarian who has been trained in principles of classification and information retrieval, a full time administrative officer, a contract specialist and an expert on project documentation. Their contracts are likely to be bigger, more complicated, and more numerous, and in some ways – because so many more people are involved – their need for a common framework for information is even greater.

Using CI/SfB

Documentation in the smaller office

The offices of quantity surveyors and consultants, builders and sub-contractors, like those of architects, all operate at varying levels of size and complexity. Some need a very simple framework for information and others need a lot of detail, and it is essential that a co-ordinating system like CI/SfB acknowledges this.

The manual itself inevitably contains and presents to all its users the whole of CI/SfB; and consequently caters for all varieties of use. Those who want only a simple method of classifying and filing a small number of documents for easy retrieval can use CI/SfB for this alone. Those who also want to use the same system to structure and reference normal project information can do so. Large offices or groups of offices which need a sophisticated system by which they can arrange and identify their reference and project information in detail, can use CI/SfB for many purposes. It is for users to decide at what level and for what purposes they want and need to use the system. The best general advice that can be given is **always** to use it in the simplest appropriate way, applying the smallest range of divisions which will identify information sufficiently for the purpose required. This will mean that some applications use it in greater depth than others.

The subject headings which make up the system are given in Tables 0, 1, 2, 3 and 4. Preceding the detail of each table there is a matrix which gives a birds-eye view of the whole table. To emphasise the advice that a decision must be made by any organisation as to the degree of detail which is to be used, **panels** giving only the main headings of CI/SfB introduce the

Using CI/SfB

detailed breakdown of each section of the Tables. **The headings in these panels are detailed enough for small office libraries and for most types of project information**. Many users will seldom need to use CI/SfB at greater depth than is suggested in these panels. Any inclination to involve more detailed division should be carefully considered and the real need for it established.

A simple presentation of the whole system is given on the wall chart which is sold with every copy of the manual and is also available separately.

Explanations of how to use the tables for different purposes are given in **Applications** 1 Project information and 2 General information.

1.3 To avoid unnecessary difficulties . . .

Ends matter more than means. Don't let SfB take over.

Using the system for producing information

Table 0 headings are already used for structuring books providing design information; Table 1 headings and codes are used for this, also for structuring cost plans and sets of drawings; Tables 1 and 2 are used for structuring national, office and job specifications and bills of quantities; Table 4, based on the CIB Master Lists, is used to structure product data sheets. Headings and codes from all the tables have been used for structuring sets of articles in the technical press.

Using CI/SfB

Essentially, the author or draughtsman producing information needs to:

1. Read the sections of this manual relevant to the type of document he is producing.

2. Decide broadly how the documentation will best be organised – whether relevant CI/SfB headings will be used as the primary subject division (as with an elemental bill of quantities using Table 1 elements); or as a means of subdividing other categories (as with an SMM arranged bill of quantities in which each SMM work section is divided by relevant Table 1 elements).

3. Plan the scope of each part of the documentation before producing any of it (eg preparing a preliminary drawings list for a set of working drawings, or a preliminary contents list for a code of practice or report), keeping as close as possible to CI/SfB concepts and arrangements.

4. Explain to users exactly how the information has been organised, to help them find their way easily to what they require to know.

Flexibility

Where CI/SfB is used for some private purpose like the organisation of personal files, it can be modified in any way that appears necessary. Where it is used for the arrangement of documents for contractors and others who receive information from many separate sources, there is a stronger case for using the system generally in accordance with the recommendations in this manual.

Coding may not be necessary. Many people feel obliged to use codes or even to use them in a particular way. In fact, for some applications there may be no need to use codes at all: it may be enough to use the headings of the system. Where codes are used, common sense should decide their form. If (6) is preferred to (6–), or (91) instead of (90.1) these alternatives can be used. In some cases (eg, for computer applications), a fixed code length will be needed, requiring adaptations such as h0 instead of h for metals. Where codes are not used, headings may be used in alphabetical order instead of system order. Where special codes or headings are used, the person receiving information should be informed of changes that have been made from the published recommendations.

CI/SfB, like any other information aid, is merely a means to an end, not an end in itself. Above all, it should be used in a simple way appropriate to each separate purpose.

Complexity

Many users try to apply to CI/SfB in ways which are unnecessarily detailed. The aim is to provide manageable groups of information, not using more detail from CI/SfB than is needed to achieve this. Only when the amount of information at any CI/SfB heading and code seems likely to become unmanageable should the use of the more detailed breakdown of that code be considered. If a library has about ten documents on religious buildings each can simply be coded 6 from Table 0. If it has more documents than this, it may be necessary to classify them using more detailed headings and two-digit codes, eg 61, 62 etc. If there are still more, it may be necessary to classify them by Table 1 and the other tables too; but the amount of classification and coding should always be kept to a minimum.

Using CI/SfB

1.4 Basic references

CI/SfB Project manual
Organising building project information, incorporating the authoritative UK version of the international SfB classification system as it applies to project information.
Architectural Press, London 1971.
Copyright RIBA Services Ltd.
Out of print.

Construction Industry Thesaurus (CIT)
Development editions.
Department of the Environment, 1973.
Available from the Polytechnic of the South Bank, Wandsworth, London SW8 2JZ.

National Building Specification (NBS)
London 1973.
Copyright NBS Ltd.
Out of print.
Subscription service available from NBS Ltd, 6 Higham Place, Newcastle upon Tyne, NE1 8AF.

CIB Master Lists
Report No. 18, CIB Rotterdam 1972.
Available from the Building Centre,
26 Store Street, London WC1E 7BT.

The SfB System
Report No. 22, CIB Rotterdam 1973.
Authorised building classification system for use in project information and related general information.
Available from CIB General Secretariat, 704 Weena, PO Box 20704, Rotterdam, Holland.

CI/SfB Construction classification wallchart
RIBA Publications Ltd, London 1974.
Corresponds with this 1976 edition of the CI/SfB Manual.
Available from RIBA Publications Ltd, 66 Portland Place, London W1N 4AD.

Bibliography on Data Co-ordination in the Construction Industry
Department of the Environment, London.
Revised versions published periodically by the Department.

Resources control
RIBA Management Advisory Committee.
RIBA Publications Ltd, London 1976.
Available from RIBA Publications Ltd.

Libraries for professional practice
Edited by Patricia Calderhead.
Architectural Press, London 1972.

RIBA Standard printed drawing sheets
Distributed by DOME Drawing Office Materials & Equipment Ltd,
Lynch Lane, Weymouth, Dorset DT4 9DW.
For quotations and samples apply to DOME.

Tables

Introduction to the tables

This presentation of CI/SfB is followed by the five tables in detail, each preceded by a key and a matrix showing the table as a whole.

Table 0	**Physical environment**	**End results, projects, the physical environment as a whole**
	Planning areas, coded 0	eg: 051 Villages Areas comprising several of the facilities below
	Facilities, coded 1 to 8	eg: 41 Hospitals Facilities (works, buildings, internal and external spaces) for particular purposes
	Common facilities, coded 91 to 97	eg: 93 Kitchens
	Other physical features of the environment, coded 98, 99	eg: Monuments
Table 1	**Elements**	**Constructional parts of projects according to their function**
	Sites, projects, coded (––)	
	Fabric, coded (1–) to (4–)	eg: (41) Finishes externally
	Services, coded (5–) to (6–)	eg: (53) Water supply
	Fittings, coded (7–) to (8–)	eg: (73) Culinary fittings
	External works, coded (9–)	
Table 2	**Constructions, forms**	**Constructional parts of projects, according to their form**
	Formed and formless work, products, coded E to Y	eg: F Brickwork, bricks
	Joints, coded Z	
Table 3	**Materials***	**Materials**
	Building materials by substance, coded e to s	eg: h metal
	Building materials by purpose, coded t to w	eg: t fixing and jointing materials
	Substances, coded z	

Table 4	Activities, requirements	Non-objects and objects which assist or affect construction but are not incorporated in it
	Construction activities and agents, coded (A) to (D)	eg: (A) Administration
	Properties, processes and agents, coded (E) to (S)	eg: (K) Fire
	Application, users, coded (T) to (U)	Factors relating to users and their activities
	Operation, maintenance, coded (V) to (W)	eg: (W2) Cleaning
	Change, coded (X)	Factors relating to change
	Commerce, coded (Y)	Factors relating to commerce
	Other categories, coded (Z)	Matters not otherwise included

The headings given in the panels at the beginning of each main subject group in the tables are examples only. Most users will need to use part – perhaps only a small part – of the formal detailed schedules from time to time but a few may find that the panels provide as much detail as they will ever want.

The panel at (2–) suggests, for example, that an office library may need one or more files for (27) Roofs; that sets of working drawings may include dimensioned location drawings, assembly or component drawings coded (27) Roofs; and that Elemental specifications, bill of quantities and cost plans may also include items under this code and heading. The applications suggested for particular headings included in the panels are based on the successful use of CI/SfB by offices, but have no other significance or authority.

* The title of international SfB Table 3 is 'Materials and other resources' and classes a to d cover resources other than materials. In CI/SfB these are at (A) to (D) in Table 4.

Table 0

Physical environment

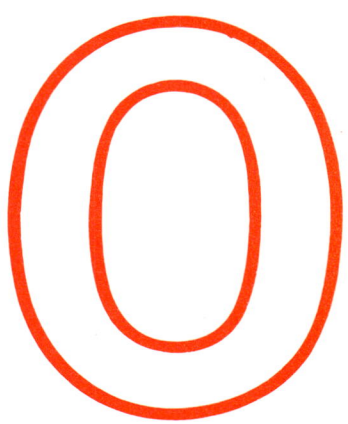

Key

Scope
End results of the construction process including planning areas, facilities such as works, buildings, internal or external spaces for particular purposes.

Structure
Three main divisions:
Areas and complexes, coded 0;
Facilities for particular purposes, coded 1 to 8;
Facilities for common activities, coded 9.
A diagrammatic view of the whole table is shown on the next page.

Changes
Most changes from the last edition are at a detailed level and aim to make the table a useful tool for the classification of information on internal and external spaces eg 'computer rooms' as well as on building types. Classes 1 and 2 have been substantially re-arranged.

Coding
1 followed by 10 to 19;
then 2 followed by 20 to 29;
and so on.
Code positions marked 'vacant' may be used for any purpose in private applications but not on published literature.

Use
Where two classes seem equally relevant, use the first in the absence of any private rule to the contrary. When using the table for a particular purpose consider:

Simplicity of use
For a small set of data it may be possible to use just the nine division codes and headings, eg 1 Utilities, and ignore all the detailed codes and headings. Use codes only if they serve a useful purpose.

Compound subjects
Adopt suitable rules for dealing consistently with documents or items which relate to two or more headings.
In cases where a very detailed systematic breakdown is required, two codes from the table can be used together, separated by a colon eg
2:94 Sanitary spaces in industrial buildings
282:94 Sanitary spaces in factories

Filing
The heading 'Other types etc' at the end of each class indicates that all other aspects of the main subject not listed specifically may follow on at the position.
Some users may prefer to ignore these headings and file this material immediately after the main heading (see Applications 2.3).

Table 0 matrix

Planning areas		Facilities							
0	**Planning areas**	**1**	**Utilities, civil engineering facilities**	**2**	**Industrial facilities**	**3**	**Administrative, commercial, protective service facilities**	**4**	**Health, welfare facilities**
00	Vacant	10	Vacant	20	Vacant	30	Vacant	40	Vacant
01	Vacant	11	Rail transport facilities	21	Vacant	31	Official administrative facilities, law courts	41	Hospital facilities, hospitals
02	International, national planning areas	12	Road transport facilities	22	Vacant	32	Office facilities, offices	42	Other medical facilities
03	Regional, sub-regional planning areas	13	Water transport facilities	23	Vacant	33	Commercial facilities	43	Vacant
04	Vacant	14	Air transport facilities, other transport facilities	24	Vacant	34	Trading facilities, shops	44	Welfare facilities, homes
05	Rural, urban planning areas	15	Communications facilities	25	Vacant	35	Vacant	45	Vacant
06	Land use planning areas	16	Power supply, mineral supply facilities	26	Agricultural facilities	36	Vacant	46	Animal welfare facilities
07	Vacant	17	Water supply, disposal facilities	27	Manufacturing facilities	37	Protective service facilities	47	Vacant
08	Other planning areas	18	Other utilities, civil engineering facilities	28	Other industrial facilities	38	Other administrative, commercial, protective service facilities	48	Other health, welfare facilities
09	Common areas relevant to planning	19	Vacant	29	Vacant	39	Vacant	49	Vacant

Table 0 matrix

5	Recreational facilities	6	Religious facilities	7	Educational, scientific, information facilities	8	Residential facilities	9	Common facilities, other facilities
50	Vacant	60	Vacant	70	Vacant	80	Vacant	90	Vacant
51	Refreshment facilities	61	Religious centre facilities	71	Schools facilities	81	Housing	91	Circulation, assembly facilities
52	Entertainment facilities	62	Cathedrals	72	Universities, colleges, other education facilities	82	One-off housing units, houses	92	Rest, work facilities
53	Social recreation facilities	63	Churches, chapels	73	Scientific facilities	83	Vacant	93	Culinary facilities
54	Aquatic sports facilities	64	Mission halls, meeting houses	74	Vacant	84	Special housing facilities	94	Sanitary, hygiene facilities
55	Vacant	65	Temples, mosques, synagogues	75	Exhibition, display facilities	85	Communal residential facilities	95	Cleaning, maintenance facilities
56	Sports facilities	66	Convents	76	Information facilities, libraries	86	Historical residential facilities	96	Storage facilities
57	Vacant	67	Funerary facilities, shrines	77	Vacant	87	Temporary, mobile residential facilities	97	Processing, plant, control facilities
58	Other recreational facilities	68	Other religious facilities	78	Other educational, scientific, information facilities	88	Other residential facilities	98	Other types of facilities, buildings
59	Vacant	69	Vacant	79	Vacant	89	Vacant	99	Parts of facilities, other aspects of the physical environment, architecture as a fine art

0   Planning areas

0   PLANNING AREAS

If subdivision of this class is needed it will often be enough to use only the headings below:
- **02 International, national planning areas**
- **03 Regional, sub-regional planning areas**
- **05 Rural, urban planning areas**
- **06 Land use planning areas**
- **08 Other planning areas**
- **09 Common areas**

The schedule may be used in a flexible way to divide information into eg three classes:
- 02/03 International/national, regional planning areas
- 05 Rural, urban planning areas
- 06/09 Land use planning areas; other areas

00 Vacant

01 Vacant

02 INTERNATIONAL, NATIONAL PLANNING AREAS
Problems and projects relating to the atmosphere, surface areas including sea and land areas, underground and underwater areas, where these are international or national in scale

03 REGIONAL, SUB-REGIONAL PLANNING AREAS
Parts of national planning areas
If necessary, subdivide eg:
- 031 Regional planning eg: South East region

Sub-regional planning areas 033/038
- 033 County planning, including metropolitan counties
- 035 District planning, including metropolitan districts
- 036 Local planning
- 038 Other sub-regional planning areas eg: structure plans (formerly 'development plans'), tactical plans

04 Vacant

Town and country planning 05/08

05 RURAL, URBAN PLANNING AREAS
If necessary, subdivide eg:
- 051 Rural, country planning, green belt, green wedges; Villages, hamlets

Urban planning, built-up areas 052/058
- 052 Conurbations, metropolises, city regions
- 053 Cities, towns eg: market towns, satellite towns, dormitory towns, holiday towns, industrial towns
- 054 New towns, new cities, garden cities, utopian towns
- 055 Expanding towns
- 056 Other settlements, communities eg: shanty settlements, gypsy encampments, communes
- 058 Other types of urban planning and built up areas eg: town centres, suburbs, squares, neighbourhoods, overspill areas, open spaces in built up areas, townscape, streets
  Recreational open space see 06:5

06 LAND USE PLANNING AREAS
Existing or proposed use eg:
- 06:11 Rail transport land use
- 06:261 Use for shelter belts
- 06:264 Horticultural land use
- 06:87 Caravan sites

07 Vacant

08 OTHER PLANNING AREAS
If necessary, subdivide eg:
Development planning areas 081/083
- 081 Conservation, restoration, improvement planning areas; designated areas
- 082 Large scale development areas eg: redevelopment areas
- 083 Other development planning areas
  Statutory designated areas 0831/0832
  1 Areas scheduled for intense development
  2 Areas scheduled for renewal, action areas (formerly 'comprehensive development areas')
  3 Twilight areas, depressed areas, declining areas, slums
  4 Derelict areas, waste land
  8 Other types, parts of development areas

- 084 Planning areas and works by form eg: linear = ribbon, cluster, concentric growth, gridiron
- 085 Infill areas, high density areas
  Infill buildings see 982
- 087 Natural areas in the service of man
  1 National parks, commons
  2 Areas of special scientific interest
  3 Areas of outstanding natural beauty
  4 Areas rich in natural flora and fauna
  8 Other types, parts of natural areas
  Areas in the landscape see 092
  Play, sports, outdoor recreation, holiday facilities see 5
  External spaces, gardens see 998
  Environmental features see 998
- 088 Other types, parts of planning areas

09 COMMON AREAS
Pervasive areas common to two or more of classes 02/08
If necessary, subdivide eg:
- 091 Atmospheric areas
- 092 Surface areas, areas in the landscape
  Land areas eg: coastal areas, riverside areas, mountainous areas, marshy areas, hilly areas, valleys, cliffs, beaches, islands
- 093 Water areas
  Sea areas eg: underwater areas, seabed areas
  Estuaries, deltas, rivers, streams, lakes, pools, ponds
- 094 Underground eg: caves
  Ground under water areas
- 095 Areas by natural environmental factors eg: low rainfall areas, temperate areas, volcanic areas, high altitude areas
- 096 Areas by economic environmental factors eg: areas with particular types of ownership, tenanted areas, highly-priced areas
- 097 Areas by social environmental factors eg: racially mixed areas, high amenity areas, areas with particular types of population, areas administered in various ways
- 098 Areas by other aspects of town and country planning

1 Utilities, civil engineering facilities

1 UTILITIES, CIVIL ENGINEERING FACILITIES

Includes transport, telecommunications, basic supply and disposal facilities, general engineering works.

If subdivision of this class is needed it will often be enough to use only the headings below:

11	**Rail transport facilities**
12	**Road transport facilities**
13	**Water transport facilities**
14	**Air transport facilities, other transport facilities**
15	**Communications facilities**
16	**Power supply, mineral supply facilities**
17	**Water supply; disposal facilities**
18	**Other utilities, civil engineering facilities**

The schedule may be used in a flexible way to divide information into eg three classes:
- 11/14 Transport facilities
- 15 Communications facilities
- 16/18 Supply and disposal facilities etc

---

10 Vacant

Transport and communications facilities 11/15
Transport facilities 11/14
Land transport facilities 11/12

11 RAIL TRANSPORT FACILITIES
Includes railway engineering, rail transportation
If necessary, subdivide eg:

- 111 Surface, overhead railways eg: intercity, suburban lines
  Stations if shown separately see 114
- 112 Underground railways
- 113 Other railways (track operated transport) eg: mountain, rack, cable railways; monorails; guideways; cable transport including cable ways, aerial ropeways eg: cable cars, ski-lifts
- 114 Embarkation facilities eg: stations, terminals, stops, halts, platforms
- 116 Rail vehicle control facilities eg: signal boxes, marshalling yards
- 117 Rail vehicle storage, repair facilities eg: loco sheds, carriage sheds
- 118 Other types, parts of rail transport facilities eg: permanent way, track; rail bridges, rail tunnels; (rolling stock may also be included here)

12 ROAD TRANSPORT FACILITIES
Includes highway engineering, road transportation including bus systems, road haulage systems
If necessary subdivide eg:

- 121 Motorways, autobahns
- 122 Other motor roads
  1. Primary roads other than motorways (trunk roads, arterial roads, main roads)
  2. Secondary roads (minor roads)
  3. By-passes (loop roads), ring roads, radial roads
  4. Access roads, drives, approach roads, culs-de-sac
  5. Single and dual carriageway roads, grade separated roads
  8. Other types, parts of motor roads
- 123 Other roads
  1. Pedestrian streets
  2. Cycle tracks
  3. Bridleways
  4. Footpaths (field paths, footways, paths), towpaths
  8. Other types
- 124 Embarkation facilities eg: coach stations, bus stations, bus stops, bus shelters
- 125 Carparks, parking bays, street parking, loading bays
- 126 Road vehicle control facilities eg: filling stations (petrol stations), traffic controls
- 127 Road vehicle storage, repair facilities eg: garages, spare parts and accessories shops, showrooms, repair shops; inspection pits, washbays, degreasing, lubricating, body building units
- 128 Other types, parts of road transport facilities eg: carriage ways including traffic lanes, junctions (intersections); slip ways, laybys (drive ins), passing places, skid pads; hardstandings, pavements, refuges (islands), crossings; road tunnels, road bridges, road flyovers; street furniture, hoardings; (road vehicles may also be included here)

13 WATER TRANSPORT FACILITIES
Includes sea, river, canal transport
If necessary, subdivide eg:

- 131 Maritime water transport facilities including coastal, ocean
  Harbours if shown separately see 134
- 133 Inland water transport facilities including canal, non-tidal river, lake transport
  Other types, parts of waterways
  Water flow control see 187
- 134 Embarkation facilities eg: harbours (ports) eg: hovercraft stations, transit sheds, terminals
  Customs houses see 315
  Marinas see 546
  Pier buildings see 988
- 136 Boat control facilities eg: lighthouses, beacons, lifeboat stations, coastguard stations
- 137 Boat storage, repair facilities eg: repair yards, boathouses
  Shipbuilding, boatbuilding facilities see 2755
- 138 Other types, parts of water transport facilities eg: locks; docks including wet, dry, floating docks (*see also 2755*); jetties (piers), quays (wharves), berths, moorings; slipways, gangways; (ships, barges may also be included here)

1 Utilities, civil engineering facilities

14 AIR TRANSPORT FACILITIES, OTHER TRANSPORT FACILITIES
If necessary, subdivide eg:
- **141** Airport facilities including aerodromes (air fields)
  Terminals if shown separately see 144
- **142** Heliports, VTOL facilities
- **144** Embarkation facilities eg: terminals
- **146** Aircraft control facilities eg: air traffic control tower, radar facilities
- **147** Aircraft storage, repair facilities eg: hangars
- **148** Other types, parts of air transport facilities eg: landing pads, runways, taxi ways, stop ways, clear ways, aprons, holding bays, disposal bays; (aircraft may also be included here)
- **149** This position may be used for pervasive facilities common to two or more of classes 11/14 eg: embarkation facilities, vehicle control facilities, vehicle storage and repair facilities, level crossings, transportation and travel facilities generally (air cushion transport may also be included here)

15 COMMUNICATIONS FACILITIES
If necessary, subdivide eg:
Telecommunications facilities 151/156
- **151** Broadcasting facilities
- **152** Radio facilities
- **153** Television facilities eg: closed circuit television
- **154** Telephone, telegraph facilities
  1. Telephone facilities eg: telephone exchanges
  2. Telegraph facilities eg: telex, facsimile transmission, data communications
- **156** Lines facilities; micro-wave beam, laser beam, radio wave facilities
  Transmission facilities eg: transmitting, receiving, monitoring stations including satellite ground stations, aerial masts and towers
  Switching facilities eg: exchanges, switching centres, telecommunications control facilities
  Other telecommunications facilities
- **157** Postal communications facilities eg: post offices, sorting offices, mail rooms, parcels offices
- **158** Other types, parts of communications facilities

16 POWER SUPPLY, MINERAL SUPPLY FACILITIES
If necessary, subdivide eg:
- **161** Heat supply facilities
  Heat production facilities eg: geothermal
  Heat dissipation facilities eg: cooling towers, freestanding chimneys
- **162** Electricity supply facilities
  power production, electricity, power stations (generating stations) including nuclear, hydroelectric;
  power transmission including substations
- **163** Mechanical power supply facilities eg: windmills, watermills
  Solar energy supply facilities
  Facilities for fuels other than fossil fuels

  Fossil fuel extraction, supply facilities 164/166
- **164** Oil (petroleum) extraction, supply facilities
  oil wells, oil storage facilities, oil pipelines, oil rigs, oil refineries
- **165** Gas extraction, supply facilities
  gas storage facilities, gas pipelines, gas rigs, gas wells
- **166** Solid fuel extraction, supply facilities eg: coal mines (collieries), pits, galleries, shafts; peat cuttings
- **167** Other mineral extraction, supply facilities eg: mines, quarries, open-cast workings
- **168** Other types, parts of mineral supply facilities

17 WATER SUPPLY; DISPOSAL FACILITIES
Public health engineering
If necessary, subdivide eg:
- **171** Water supply facilities
  1. Water supply storage facilities eg: wells, reservoirs, water tanks
  2. Water supply treatment facilities eg: water works
  3. Water supply distribution facilities eg: pumping stations
  8. Other water supply facilities

  Disposal facilities 174/178
- **174** Sewage disposal facilities
  1. Sewage collection, storage facilities eg: sewer systems, cesspools
  2. Sewage treatment facilities eg: sewage farms; sewage disposal, re-use facilities
  8. Other sewage disposal facilities
- **175** Refuse disposal facilities
  1. Refuse collection, storage facilities eg: refuse depots
  2. Refuse treatment facilities
  3. Refuse tips, dumps, re-use facilities
  8. Other refuse disposal facilities
- **176** Mineral waste disposal facilities eg: slag heaps, nuclear fission waste products disposal facilities
- **177** Disposal of the dead eg: mortuaries, morgues
  Crematoria, cemeteries etc see 67
- **178** Other types, parts of waste disposal facilities

18 OTHER UTILITIES, CIVIL ENGINEERING FACILITIES
If necessary, subdivide eg:
- **181** Tunnels, subways, underpasses, culverts, pipelines
- **182** Bridges
  1. Aqueducts
  2. Vehicle bridges
  3. Footbridges
  4. Pipe, cable bridges
  5. Movable bridges eg: horizontal including swing, roller, transporter bridges; vertical including vertical lift, bascule, counterbalanced, cable lift bridges
  6. Viaducts
  8. Other bridges eg: suspension, cantilever, arch, portal frame; cable braced (stayed) bridges
- **183** Towers and other vertical engineering structures
  Aerial masts and towers see 156
  Chimneys, cooling towers see 161
- **184** Bulk goods storage facilities
  1. Silos, bins, hoppers, tanks
  2. Structures with thrust resistant walls
  3. Structures with non-thrust resistant walls
  4. Structures with little or no wall
  8. Other bulk goods storage facilities eg: dumps
  See also 164, 165, 171, 175, 268
- **185** Land reclamation eg: polders, fens
- **186** Land retention facilities
  1. Avalanche protection
  2. Landslide protection
  3. Erosion protection eg: soil erosion, sea defence, coastal protection facilities eg: sea walls, breakwaters (moles), groynes
  4. Revetments
  8. Other land retention facilities
- **187** Water retention and flow control
  Water retention facilities eg: dykes, levees, cofferdams; dams, barrages, weirs
  Water control facilities eg: flood protection, drainage, irrigation
- **188** Other types, parts of utilities, civil engineering facilities eg: supports

19 Vacant

2 Industrial facilities

2 INDUSTRIAL FACILITIES
Includes agricultural and manufacturing facilities
Power supply, mineral supply facilities see 16
This class may apply to eg: a works, an individual block, a department, a room, an internal or external space

If subdivision of this class is needed it will often be enough to use only the headings below:
- **26 Agricultural facilities**
- **27 Manufacturing facilities**
- **28 Other industrial facilities**

20 Vacant

21 Vacant

22 Vacant

23 Vacant

24 Vacant

25 Vacant

26 AGRICULTURAL FACILITIES
If necessary, subdivide eg:
- **261** Forestry facilities, shelter belt facilities
- **262** Fishing facilities, fisheries
- **263** Farming facilities eg: farms
- **264** Horticultural (market gardening) facilities eg: nurseries, hothouses, greenhouses, glasshouses; agricultural produce facilities
- **265** Livestock facilities
  Subdivide as 464 eg:
  2654 cattle unit, cowshed, milking parlour
  Abattoirs, dairies see 273
  Animal houses generally see 46
- **268** Other types, parts of agricultural facilities eg:
  1 Silos, bins, hoppers, tanks
  2 Bulk storage with thrust-resistant walls
  3 Storage with non-thrust resistant walls
  4 Storage with little or no wall, dutch barns
  5 Controlled environment facilities
  8 Fields, yards, pens, pits

27 MANUFACTURING FACILITIES
Codes in Roman numerals on the right below are taken from the Standard Industrial Classification (SIC), available from HMSO
Factories etc for specific purposes eg:

273	Food, drink, tobacco industries eg: abattoirs, dairies	III
274	Chemicals and allied industries	V
275	Engineering industries, metal industries	
	1 Metals	VI
	2 Mechanical engineering	VII
	3 Instrument engineering	VIII
	4 Electrical engineering	IX
	5 Shipbuilding and marine engineering eg: launching ramps	X
	Ship repair if not at 137; ships if not at 138	
	6 Vehicles	XI
	Rail vehicles if not at 11; road vehicles if not at 12	
	8 Other metal goods	XII
276	Textile, clothing industries	
	1 Textiles	XIII
	2 Leather, leather goods and fur	XIV
	3 Clothes, footwear	XV
277	Clay, cement, timber industries	
	1 Bricks, pottery, glass, cement	XVI
	2 Timber, furniture	XVII
	3 Paper, printing and publishing	XVIII
	4 Other manufacturing industries	XIX
278	Construction industry eg: builders' yards	XX

28 OTHER INDUSTRIAL FACILITIES
Facilities common to two or more industries
If necessary, subdivide eg:
- **281** Heavy industry facilities eg: works, mills
- **282** Factories eg: standard factories, flatted factories, industrial workshops
- **284** Industrial storage facilities eg: industrial warehouses
- **285** Industrial process facilities eg: assembly lines, industrial work facilities, production machine rooms
- **288** Other types, parts of industrial facilities eg: repair facilities

29 Vacant

3 Administrative, commercial, protective service facilities

3 ADMINISTRATIVE, COMMERCIAL, PROTECTIVE SERVICE FACILITIES
This class may apply to eg: a multi-block project, an individual block, a department, a room, an internal or external space

> If subdivision of this class is needed it will often be enough to use only the headings below:
> 31 **Official administrative facilities, law courts**
> 32 **Office facilities, offices**
> 33 **Commercial facilities**
> 34 **Trading facilities, shops**
> 37 **Protective service facilities**
> 38 **Other administrative, commercial, protective service facilities**
>
> The schedule may be used in a flexible way to divide information into eg three classes:
> 31/32 Official and office facilities
> 33/34 Commercial and trading facilities
> 37/38 Protective service facilities etc

30 Vacant

31 OFFICIAL ADMINISTRATIVE FACILITIES, LAW COURTS
Offices see 32
If necessary, subdivide eg:
311 International legislative and administrative facilities eg: European parliament, EEC Commission, UN
312 National legislative and administrative facilities eg: parliaments, capitols, ministries
314 Regional and local legislative and administrative facilities eg: regional, county and district offices; civic centres; county, city and town halls; guildhalls; mayor's parlour
315 Local offices of government departments eg: taxation facilities eg: customs houses; labour exchange facilities eg: employment exchanges, job centres
316 Official representation facilities eg: palaces, presidential residences, embassies, consulates, legations, high commissions
317 Law courts eg: civil courts including county courts, magistrates courts; criminal courts, assizes; court rooms, bench, dock, witness box, jury box
318 Other types, parts of official administrative facilities eg: ceremonial suites, robing rooms, debating chambers

32 OFFICE FACILITIES, OFFICES
Offices associated with a particular facility see that facility eg: shops and offices see 34
Drawing offices, design offices, studios; art, photographic studios; professional offices, executive offices, offices of nationalised industries, secretarial offices
Landscaped offices (Bürolandschaft)
Open plan spaces generally see 997

33 COMMERCIAL FACILITIES
If necessary, subdivide eg:
331 Mixed commercial developments
335 Insurance facilities eg: underwriting
336 Building societies
337 Investment facilities eg: stock exchanges, stockbroking facilities
338 Other types, parts of commercial facilities eg: banks including banking halls, tellers boxes, safe deposits

34 TRADING FACILITIES, SHOPS
If necessary, subdivide eg:
341 Wholesaling facilities, auction rooms; bulk-buying stores, discount trading stores, mail order stores, cash and carry stores
342 Shopping centres including local centres, shopping arcades, markets
343 Department stores
344 Hypermarkets, supermarkets
345 Self-service shops, corner shops, shops by goods sold eg: food shops, durable goods shops
346 Service shops eg: dry cleaners, cobblers, launderettes, forges, gunsmiths, upholsterers, book binders, tailors
348 Dealers eg: coal merchants, builders' merchants
Other types, parts of trading facilities eg: shop showrooms, fitting facilities, sales facilities including booking halls, trading stalls, kiosks

35 Vacant

36 Vacant

37 PROTECTIVE SERVICE FACILITIES
If necessary, subdivide eg:
371 Coastguard, lifeboat stations (if not at 136)
372 Fire protection facilities eg: fire stations, fire practice towers, smoke chambers, fire hose drying spaces
373 Ambulance facilities
374 Civil law, criminal law enforcement facilities eg: police stations
375 Defence facilities
Armed forces facilities 3751/3753
 1 Air force facilities
 2 Navy facilities
 3 Army facilities, other armed forces facilities eg: barracks generally
 4 Civil defence facilities
 5 Camps, depots, bases, ranges
 6 Blockhouses, city walls
 7 Air raid shelters, fall-out (radiation) shelters
 8 Aggressive facilities eg: missile sites;
 Other defence facilities
376 Prisons (gaols, jails, penitentiaries)
 1 Open prisons
 2 Closed prisons eg: semi-secure prisons; secure prisons; maximum security prisons
 5 Reformatories (borstals)
 8 Other detention facilities eg: assessment centres
Approved schools see 7175
378 Other types, parts of protective service facilities eg: secure facilities

38 OTHER ADMINISTRATIVE, COMMERCIAL, PROTECTIVE SERVICE FACILITIES

39 Vacant

4 HEALTH, WELFARE FACILITIES

This class may apply to eg: a multi-block project, an individual block, a department, a room, an internal or external space

> If subdivision of this class is needed it will often be enough to use only the headings below:
> **41 Hospital facilities, hospitals**
> **42 Other medical facilities**
> **44 Welfare facilities, homes**
> **46 Animal welfare facilities**
> **48 Other health, welfare facilities**

40 Vacant

Health, medical facilities 41/42

41 HOSPITAL FACILITIES, HOSPITALS
If necessary, subdivide eg:
- 411 Teaching hospitals including postgraduate teaching centres
- 412 General hospitals, GP hospitals, cottage hospitals

Hospital facilities by part of body 413/414
- 413 Mental, psychiatric
- 414 Ear, nose and throat, dental, heart, other types by part of body

Hospital facilities by patient 415/416
- 415 Maternity, gynaecological
- 416 Paediatric (children); geriatric (old people)
- 417 Hospital facilities by technique
  1. Diagnosis including radiography (X-ray)
  2. Surgery including operating theatres
  4. Pathology
  5. Physical medicine, occupational therapy
  7. Chemotherapy, including pharmacies, dispensaries
  8. Other techniques including physiotherapy, radio therapy
- 418 Other types, parts of hospital facilities eg:
  outpatients facilities, casualty facilities; day patients facilities; in-patients facilities; admission units; private patients facilities; isolation facilities; field hospital facilities; nursing facilities including wards

42 OTHER MEDICAL FACILITIES
If necessary, subdivide eg:
- 421 Health centres; health clubs
- 422 Clinics eg:
  maternity and child clinics, geriatric, screening clinics
- 423 Surgeries including group practices, doctors' surgeries, consulting rooms
- 424 Special centres, clinics, surgeries eg:
  dentists' surgeries
- 426 First aid posts including emergency and field posts
- 427 Medical research facilities
- 428 Other types, parts of medical facilities eg:
  blood transfusion, forensic facilities, examination facilities

43 Vacant

44 WELFARE FACILITIES, HOMES
If necessary, subdivide eg:
- 442 Nursing homes, convalescent homes, sanatoria

Welfare facilities by condition of user 443/445
- 443 Chronic invalids, addicts
- 444 Mentally handicapped
- 445 Physically handicapped eg:
  polio, spastic, blind, deaf;
  Other welfare facilities by condition of user eg:
  unmarried mothers, battered wives, destitute people

Welfare facilities by age of user 446/447
- 446 Orphanages, nurseries (crèches)
- 447 Old people's homes
  Old people's housing generally see 843
  Other welfare facilities by age of user
- 448 Other types, parts of welfare facilities eg:
  overnight, short stay, long stay accommodation

45 Vacant

46 ANIMAL WELFARE FACILITIES
If necessary, subdivide eg:
- 461 Veterinary hospitals
- 462 Animal clinics, dispensaries
- 463 Animal clipping and pedicuring facilities
- 464 Animal rearing and living facilities
  Agricultural facilities see 26
  1. Fish
  2. Cats, dogs (kennels)
  3. Horses (stables)
     Riding schools see 565
  4. Cattle
  5. Sheep, goats
  6. Pigs
  7. Birds, poultry
  8. Other animal rearing and living facilities eg: rodents, bees, insects, exotic animals
- 468 Other types, parts of animal welfare facilities eg:
  quarantine facilities

47 Vacant

48 OTHER HEALTH, WELFARE FACILITIES

49 Vacant

5 Recreational facilities

5 RECREATIONAL FACILITIES

Includes social recreation and refreshment facilities eg: leisure facilities

This class may apply to eg: a multi-block project, an individual block, a department, a room, an internal or external space

If subdivision of this class is needed it will often be enough to use only the headings below:
- **51 Refreshment facilities**
- **52 Entertainment facilities**
- **53 Social recreation facilities**
- **54 Aquatic sports facilities**
- **56 Sports facilities**
- **58 Other recreational facilities**

The schedule may be used in a flexible way to divide information into eg two classes:
- 51/53 Refreshment, entertainment, social recreation facilities
- 54/58 Water and other sports facilities etc

50 Vacant

51 REFRESHMENT FACILITIES

Social refreshment
Eating, culinary facilities if described separately see 93
If necessary, subdivide eg:
- 511 Canteens, refectories
- 512 Restaurants
- 515 Cafes, snack bars, coffee bars, milk bars
- 517 Public houses, bars, taverns, beer gardens, tap rooms, licensed premises
- 518 Other types, parts of refreshment facilities eg: banqueting rooms

52 ENTERTAINMENT FACILITIES

Performing arts
If necessary, subdivide eg:
- 521 Entertainment arenas; dance halls, ballrooms, discotheques
- 522 Musical facilities eg: concert halls, bandstands, music practice rooms
- 523 Operatic facilities eg: opera houses
- 524 Drama facilities eg: theatres, drama stages, fly galleries, scenery docks, property spaces
- 525 Cinemas
  Projection facilities, audio-visual facilities generally see 768
- 526 Circuses, circus rings
- 528 Other types, parts of entertainment facilities eg: stages, auditoria (audience spaces); rehearsal facilities, practice facilities; dressing, changing facilities including green rooms; entertainments production studios

53 SOCIAL RECREATION FACILITIES

Congress and conference halls see 916
If necessary, subdivide eg:
- 532 Community centres eg: cultural centres, arts centres, village halls
- 534 Clubs eg: youth centres, students unions, other non-residential, non-commercial clubs
- 536 Residential clubs
- 537 Commercial clubs, night clubs
- 538 Other types, parts of social recreation facilities

54 AQUATIC SPORTS FACILITIES

If necessary, subdivide eg:
Swimming facilities 541/544
- 541 Covered swimming pools
- 543 Open air swimming pools, lidos
- 544 Other types, parts of swimming facilities eg: diving pools
- 546 Boating facilities eg: sailing, canoeing, rowing facilities, marinas
  Harbours see 134
- 548 Other types, parts of aquatic sports facilities eg: water ski-ing facilities

55 Vacant

56 SPORTS FACILITIES

If necessary, subdivide eg:
- 561 Sports centres eg: Crystal Palace
- 562 Sports halls
  Gymnasia, physical training facilities eg: gymnastics, acrobatics, bodybuilding, weight lifting;
  Fighting sports facilities eg: boxing, fencing, wrestling, judo, aikido, karate;
  One-to-one sports facilities eg: badminton, fives, squash, tennis, billiards, snooker, table tennis
- 563 Bowling alleys (tenpin bowling alleys)
- 564 Stadia
  Athletics facilities eg: for track, field events;
  Racing facilities eg: dog, horse, cycle, motor cycle, motor; wheel sports facilities;
  Team ball games facilities eg: cricket, football, rugby, hockey, baseball, netball
- 565 Equestrian facilities eg: riding schools, polo
  Stables see 4643
  Shooting facilities eg: archery ranges
  Hunting, fishing facilities
- 566 Bowling greens, golf courses
- 567 Air sports facilities eg: gliding, aeronautics
  Winter sports facilities eg: ski schools, jumps and slopes, toboggan runs, ice rinks, curling, skating
- 568 Other types, parts of sports facilities eg: pentathlon facilities; outdoor sports facilities; grandstands (stands), pavilions, arenas, courses, tracks, pitches, courts, playing fields; sports practice, training facilities

57 Vacant

58 OTHER RECREATIONAL FACILITIES

If necessary, subdivide eg:
- 581 Gambling facilities, casinos
- 582 Amusement arcades
- 583 Fairgrounds, funfairs, showgrounds
- 585 Play facilities eg: playgrounds, adventure playgrounds, play centres, play spaces
- 587 Parks, leisure gardens eg: gardens for the blind; outdoor recreation facilities
  Natural areas in the service of man see 087
  Landscape areas see 092
  Outdoor sports facilities see 568
  Gardens, external spaces generally see 998
  Environmental features see 998
- 588 Other types, parts of recreational facilities eg: holiday facilities, holiday camps

59 Vacant

6

6 Religious facilities

6 RELIGIOUS FACILITIES
This class may apply to eg: a multi-block project, an individual block, a department, a room, an internal or external space

If subdivision of this class is needed it will usually be enough to use the headings below: no further subdivision is provided but the scope of each heading is illustrated on the right

61 Religious centre facilities
62 Cathedrals
63 Churches, chapels
64 Mission halls, meeting houses
65 Temples, mosques, synagogues
66 Convents
67 Funerary facilities, shrines
68 Other religious facilities

60 Vacant

61 RELIGIOUS CENTRE FACILITIES
eg: episcopal palaces, deaneries, pastoral centres, ecumenical centres

62 CATHEDRALS
eg: chapter houses, cathedral treasuries

63 CHURCHES, CHAPELS
eg: churches, church halls (if also used or described with churches), chapels, chantries

64 MISSION HALLS, MEETING HOUSES

65 TEMPLES, MOSQUES, SYNAGOGUES

66 CONVENTS
Residential religious facilities eg: monasteries, nunneries, abbeys, priories, friaries, retreats

67 FUNERARY FACILITIES, SHRINES
eg: crematoria, cemeteries (grave yards); funeral vaults; tombs, mausoleums; shrines, feretory spaces, relics spaces, reliquaries
Monuments see 984
Morgues, mortuaries see 177

68 OTHER RELIGIOUS FACILITIES
Chancels including sanctuaries, choirs, altars; transepts; crypts; watching galleries; belfries; side chapels; baptisteries; confessing facilities; preaching facilities; sacramental facilities; ambulatory facilities including cloisters, aisles, retrochoirs; vestries, sacristies

69 Vacant

7 Educational, scientific, information facilities

7 EDUCATIONAL, SCIENTIFIC, INFORMATION FACILITIES

Facilities for the acquisition of knowledge
This class may apply to eg: a multi-block project, an individual block, a department, a room, an internal or external space

---

If subdivision of this class is needed it will often be enough to use only the headings below:

- **71** Schools facilities
- **72** Universities, colleges, other education facilities
- **73** Scientific facilities
- **75** Exhibition, display facilities
- **76** Information facilities, libraries
- **78** Other educational, scientific, information facilities

The schedule may be used in a flexible way to divide information into eg two classes:
- 71/73 Educational, scientific facilities
- 75/78 Exhibition, information facilities etc

---

70 Vacant

Education facilities 71/72

71 SCHOOLS FACILITIES
If necessary, subdivide eg:
- 711 Nursery schools, classrooms
  Nurseries (crèches) see 446
- 712 Primary schools eg:
  infants schools, classrooms; first schools; junior (preparatory) schools; middle schools
- 713 Secondary schools (high schools) eg:
  grammar, secondary modern, secondary technical, comprehensive, community schools
- 714 Sixth form colleges, centres
- 717 Special schools eg:
  2 Educationally subnormal (ESN) schools, including severely mentally handicapped, maladjusted, delicate, hospital schools
  3 Physically handicapped
  5 Approved schools
  8 Other special schools eg: boarding, residential schools
- 718 Other types, parts of school facilities eg:
  state, independent schools; school learning, teaching facilities eg: school classrooms

72 UNIVERSITIES, COLLEGES, OTHER EDUCATION FACILITIES
If necessary subdivide eg:
- 721 Universities including technological universities, university colleges
- 722 Polytechnics, technical colleges, colleges of education, colleges of further education, colleges of art and design
- 724 Academies eg:
  music academies
- 725 Other specialist colleges
- 727 Adult education facilities; Learned societies
- 728 Other types, parts of higher education facilities eg:
  learning, teaching facilities
  Students unions see 534
  Halls of residence see 856
- 729 Other types, parts of education facilities
  Facilities common to 71/72 eg:
  1 Classrooms, lecture theatres
  2 Special subject facilities
  4 Special technique facilities eg: team teaching facilities
  8 Other facilities for imparting knowledge

73 SCIENTIFIC FACILITIES
If necessary, subdivide eg:
- 731 Research facilities
  Medical research facilities see 427
- 732 Laboratory facilities eg:
  fume chambers, instrument rooms, dark rooms
- 737 Observatories, recording stations eg:
  meteorological, geophysical, seismographic stations (earthquake stations)
- 738 Other types, parts of scientific facilities

74 Vacant

75 EXHIBITION, DISPLAY FACILITIES
If necessary, subdivide eg:
- 751 Botanical gardens; herbaria; zoos (zoological gardens)
  Animal welfare facilities see 46
- 753 Aviaries
- 755 Aquaria
- 756 Museums, planetariums
- 757 Art galleries, facilities for special display eg:
  of paintings, murals, sculptures
- 758 Other types, parts of exhibition, display facilities eg:
  design centres, building centres, exhibitions, galleries, showrooms

76 INFORMATION FACILITIES, LIBRARIES
If necessary, subdivide eg:
- 761 National libraries
- 762 Public libraries including commercial lending libraries
- 764 Information facilities by special subject eg:
  architecture
- 765 Information facilities by form of material eg:
  resource centres, illustrations, drawings, photographs
- 766 Data processing facilities eg:
  computer, reprographic facilities
- 767 Record offices, archives, patent offices
- 768 Other types, parts of information facilities eg:
  enquiry, browsing, lending, reference facilities; study facilities eg: carrels; audio-visual presentation, projection facilities

77 Vacant

78 OTHER EDUCATIONAL, SCIENTIFIC, INFORMATION FACILITIES

79 Vacant

8

8 Residential facilities

8 RESIDENTIAL FACILITIES
This class may apply to eg: a multi-block project, an individual block, a residential unit, an internal or external space

If subdivision of this class is needed it will often be enough to use only the headings below:
- 81 Housing
- 82 One-off housing units, houses
- 84 Special housing facilities
- 85 Communal residential facilities
- 86 Historical residential facilities
- 87 Temporary, mobile residential facilities
- 88 Other residential facilities

80 Vacant

81 HOUSING
Housing units (dwellings) for single families, and single family housing as a whole (including one off houses 82 and special housing 84)
If necessary, subdivide eg:
- 811 Single storey
- 812 Two storey
- 814 Three and four storey
- 815 Five storeys and over
- 816 Flats (apartments)
- 817 Maisonettes
- 818 Other types, parts of housing facilities eg:
  existing housing; housing with offices or shops; local authority, housing association, private, speculative housing

82 ONE-OFF HOUSING UNITS, HOUSES
Individually designed houses

83 Vacant

84 SPECIAL HOUSING FACILITIES
Dwellings for special classes of user.
Housing related to facilities already listed may be included at 848 or kept with them as required eg: armed forces housing see 375
If necessary, subdivide eg:
- 841 Caretakers, wardens' houses
- 843 Old people's housing including sheltered housing
  Old people's homes see 447
- 847 Single persons' housing
- 848 Other types, parts of special housing units eg:
  disabled, handicapped persons' housing

85 COMMUNAL RESIDENTIAL FACILITIES
Shared facilities
If necessary, subdivide eg:
- 852 Hotels
- 853 Motels
- 854 Guesthouses
- 856 Hostels eg:
  YMCA, youth hostels; halls of residence
- 858 Other types, parts of communal residential facilities

86 HISTORICAL RESIDENTIAL FACILITIES
Castles, chateaux, keeps, fortified houses, town houses (historical)

87 TEMPORARY, MOBILE RESIDENTIAL FACILITIES
Temporary housing units, 'half way housing' for families being re-housed; house boats; caravans (residential)

88 OTHER RESIDENTIAL FACILITIES
Living facilities eg:
living rooms
See also 92

89 Vacant

The following subdivision by form can be applied to 81 to 87
1 Detached eg:
  8111 Detached single storey housing
2 Semi-detached
4 Terraced, row houses
5 Others

9 Common facilities, other facilities

9 COMMON FACILITIES, OTHER FACILITIES

Facilities common to two or more of classes 2/8, for general activities eg: work, rest, personal hygiene, including also those in Class 1 when they occur as subordinate parts of construction work (see 914/915, 96/97) This class may apply to eg: a multi-block project, an individual block, a department, a room, an internal or external space

---

If subdivision of this class is needed it will often be enough to use only the headings below:
- **91** Circulation, assembly facilities
- **92** Rest, work facilities
- **93** Culinary facilities
- **94** Sanitary, hygiene facilities
- **95** Cleaning, maintenance facilities
- **96** Storage facilities
- **97** Processing, plant, control facilities
- **98** Other types of facilities, buildings
- **99** Parts of facilities, other aspects of the physical environment, architecture as a fine art

The schedule may be used in a flexible way to divide information into eg two classes:
- 91/97 Common facilities
- 98/99 Buildings, architecture, landscape design

---

90 Vacant

91 CIRCULATION, ASSEMBLY FACILITIES
If necessary, subdivide eg:
- 913 Access, egress facilities eg: porches, entrance halls, exits, vestibules, lobbies, reception spaces, foyers, waiting facilities
- 914 Movement facilities
  1 Horizontal circulation eg: concourses, precincts; passages (alleys) eg: corridors; links eg: link buildings, gangways, galleries, arcades
  2 Vertical circulation eg: stairways, stair enclosures; escalator ways, escalator enclosures; lift enclosures, lift lobbies
- 915 Materials handling (movement) facilities
  Large scale facilities see 11/14
- 916 Assembly facilities eg: congress, conference, assembly centres, halls
  Halls as part of churches see 63
- 917 Special assembly facilities eg: board rooms, committee rooms
- 918 Other types, parts of assembly, circulation facilities

92 REST, WORK FACILITIES
- 921 Rest facilities including rest/work facilities eg: study-bedrooms
  Sleep facilities eg: dormitories bedrooms, bunk rooms
  Living facilities eg: living rooms if not at 88
- 922 Relaxation facilities eg: lounges, break facilities, restrooms (staff rooms, common rooms), recovery rooms
- 923 Work facilities, work places, work stations
- 926 Facilities for particular skills eg: art design, photography
  Offices, studios see 32
  Dark rooms see 732
- 928 Other types, parts of rest, work facilities eg: reading, discussion, hobby facilities; consulting, interview facilities

93 CULINARY FACILITIES
Includes catering facilities
If necessary, subdivide eg:
- 931 Culinary facilities eg: kitchens
- 932 Washing-up facilities
- 934 Food preparation, cooking facilities
- 935 Culinary storage facilities eg: larders, wine cellars, pantries
- 937 Eating (dining) facilities
  Social refreshment facilities, canteens, refectories see 51
- 938 Other types, parts of culinary facilities

94 SANITARY, HYGIENE FACILITIES
If necessary, subdivide eg:
- 941 Lavatories (toilets, conveniences, latrines) including public facilities
- 942 Bathrooms, bath houses, slipper baths, sauna baths, Turkish baths; shower facilities, showers; washrooms, ablutions (as spaces)
  Sanitary fittings see (74)
- 944 Water closets, urinals (as spaces)
  Sanitary fittings see (74)
- 947 Dressing, changing rooms
- 948 Other types, parts of sanitary, hygiene facilities

95 CLEANING, MAINTENANCE FACILITIES
If necessary, subdivide eg:
- 951 Washing facilities eg: sluice rooms
- 952 Laundries (wash houses)
  Launderettes see 346
- 953 Ironing, drying, airing facilities
- 955 Utility rooms
- 958 Other types, parts of cleaning, maintenance facilities

96 STORAGE FACILITIES
See also 164, 165, 171, 175, 184, 268
If necessary, subdivide eg:
- 961 Cloakrooms, luggage rooms, stock rooms
- 963 Sheds eg: garages, car ports, car spaces, cycle sheds
  Large scale car parks etc see 125
- 964 Liquids storage facilities
- 965 Cold storage facilities (as spaces) eg: coldrooms, refrigerated storage, deep freeze storage
  Culinary cold storage fittings see (73.5)
- 966 Hot storage facilities (as spaces)
  Culinary hot storage fittings see (73.5)
- 967 Secure storage facilities eg: strong rooms (vaults)
- 968 Other types, parts of storage facilities

9   9 Common facilities, other facilities

97 PROCESSING, PLANT, CONTROL FACILITIES
Including facilities for all activities not listed or implied above eg:
plant rooms, control rooms
If necessary, subdivide eg:
- 971 Power supply eg:
  electricity
  Heat supply facilities eg:
  boiler houses
  Cold supply, chilling facilities
  Large scale facilities see 16
- 972 Water supply facilities
  Large scale facilities see 17
- 973 Waste disposal facilities eg:
  bin pounds
  Large scale facilities see 17
- 975 Control and communications facilities eg:
  control rooms
- 978 Other types, parts of processing facilities

98 OTHER TYPES OF FACILITIES, BUILDINGS
Facilities by characteristics other than purpose
Buildings generally
If necessary subdivide eg:
- 981 High rise buildings eg:
  multi-storey buildings,
  tall buildings, skyscrapers, masts, poles
  Medium rise, low rise buildings
- 982 Detached buildings; semi-detached buildings; linked buildings, including lean-to; terraces, parades; infill buildings
- 983 Demountable, floating, temporary, mobile buildings
- 984 Monuments eg:
  obelisks, cenotaphs, memorials;
  Ornamental buildings eg:
  specially decorative;
  Sculptures, murals, fountains
  Shrines see 67
- 986 Historical buildings eg:
  modern, ancient monuments
  (divide by style/date)
- 987 Buildings by particular architects, engineers (divide by name);
  The general concept of shelter
- 988 Other types of facilities eg:
  buildings in special locations eg:
  pier buildings, underground buildings; windowless buildings;
  ruinous, dilapidated, dangerous, defective buildings;
  facilities for types of users eg:
  pedestrian facilities, facilities for the handicapped

99 PARTS OF FACILITIES; OTHER ASPECTS OF THE PHYSICAL ENVIRONMENT; ARCHITECTURE AS A FINE ART
If necessary, subdivide eg:
- 992 Blocks; compartments (fire divisions); departments (functional divisions)
- 993 Storeys (horizontal divisions) eg:
  first floor, second floor, roof top
  eg: roof gardens; basements, cellars, attics (lofts, garrets), mezzanines
- 994 Wings, bays, core (vertical divisions); rooms, including halls, cubicles, booths
- 995 Facades, external aspects
- 996 Covered spaces eg:
  loggias, verandahs, conservatories, pavilions
- 997 Spaces
  Internal spaces, interior design eg:
  open plan spaces, landscaped spaces
- 998 External spaces, landscape design eg:
  landscaped spaces, gardens, spaces between buildings, specific external spaces eg:
  'areas', courtyards, patios, quadrangles, forecourts
  Urban squares, streets see 058
  Landscape see 092
  Environmental features, natural or man-made, other than those listed as facilities above eg:
  water features
- 999 Architecture as a fine art
  Any or all the subdivisions 984/988 may alternatively be placed here

Table 1

Elements

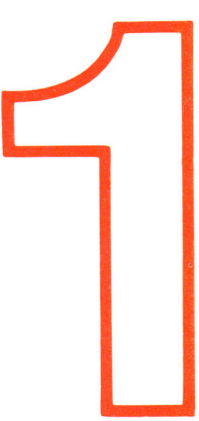

Key

Scope
Parts with particular functions which combine to make the facilities in Table 0.

Structure
Three main divisions:
The building fabric at (1–) to (4–);
Services coded (5–) tp (6–);
Fittings coded (7–) to (8–).
A diagrammatic view of the whole table is shown on the next page.

Changes
Most changes from the last edition are at a detailed level. The list of common element parts in Appendix 1 can be used for the subdivision of very large sets of data on Table 1 elements but is **not** for use for CI/SfB references in the top right hand corner on published trade literature, or for normal project applications.

Coding
(1–) followed by (10) to (19);
then (2–) followed by (20) to (29);
and so on.
Code positions marked 'vacant' may be used for any purpose in private applications but not on published literature.

Use
Where two classes seem equally relevant, use the first in the absence of any private rule to the contrary. When using the table for a particular purpose, consider:

Simplicity of use
For a small set of data it may be possible to use the nine element divisions eg: (1–) Substructure, and to ignore all the detailed codes and headings. Use codes only if they serve a useful purpose.

Compound subjects
Adopt suitable rules for dealing consistently with documents or items which relate to two or more different elements eg: 'space heating in floor beds' or 'stair and wall junctions'. In cases where a very detailed systematic breakdown is required, two codes from the table can be used together, separated by a colon eg: (13):(56) space heating systems in relation to floor beds.

Filing
The headings 'Other types'; 'Parts, accessories etc.' at the end of each class indicate that all aspects of the main subject not specifically listed may follow on at the position. Some users may prefer to ignore these headings and file this material immediately after the main heading (see Applications 2.3).

Definitions
Two definitions are given for many Table 1 elements.
The first definition should normally be used by all producing new information required for project purposes, eg: bills of quantities, drawings (see Applications 1.4 for cost planning definitions).
The second is a broader definition, relevant to the classification of already existing literature and to some project information purposes.

External elements, External works
Elements at (1–) to (8–) are 'internal' and/or 'external', but (90) is restricted to external only. For some applications it may be useful to restrict the use of (1–) to (8–) to 'internal' only, or to use eg: (20) for 'external' and (21) to (29) for 'internal'.

Table 1 matrix

(--) Sites, projects	Substructure		Structure					
	(1-)	Ground, substructure	(2-)	Structure, primary elements, carcass	(3-)	Secondary elements, completion of structure	(4-)	Finishes to structure
	(10)	Vacant*	(20)	Vacant*	(30)	Vacant*	(40)	Vacant*
			External vertical divisions					
	(11)	Ground	(21)	Walls, external walls	(31)	Secondary elements to walls, external walls	(41)	Wall finishes, external
			Internal vertical divisions					
	(12)	Vacant*	(22)	Internal walls, partitions	(32)	Secondary elements to internal walls, partitions	(42)	Wall finishes, internal
			Internal horizontal divisions					
	(13)	Floor beds	(23)	Floors, galleries	(33)	Secondary elements to floors	(43)	Floor finishes
			Vertical circulation					
	(14)	Vacant*	(24)	Stairs, ramps	(34)	Secondary elements to stairs	(44)	Stair finishes
			Ceilings					
	(15)	Vacant*	(25)	Vacant*	(35)	Suspended ceilings	(45)	Ceiling finishes
	(16)	Retaining walls, foundations*	(26)	Vacant*	(36)	Vacant*	(46)	Vacant*
			External horizontal divisions					
	(17)	Pile foundations	(27)	Roofs	(37)	Secondary elements to roofs	(47)	Roof finishes
	(18)	Other substructure elements	(28)	Building frames, other primary elements	(38)	Other secondary elements	(48)	Other finishes to structure
	(19)	Parts, accessories etc.	(29)	Parts, accessories etc.	(39)	Parts, accessories etc.	(49)	Parts, accessories etc.
		Cost summary		Cost summary		Cost summary		Cost summary

* See Appendix 6

Table 1 matrix

Services

(5-) Services, mainly piped, ducted	(6-) Services, mainly electrical
(50) Vacant*	(60) Vacant*
(51) Vacant*	(61) Electrical supply
(52) Waste disposal, drainage	(62) Power
(53) Liquids supply	(63) Lighting
(54) Gases supply	(64) Communications
(55) Space cooling	(65) Vacant*
(56) Space heating	(66) Transport
(57) Air conditioning, ventilation	(67) Vacant*
(58) Other piped, ducted services	(68) Security, control, other services
(59) Parts, accessories etc	(69) Parts, accessories etc
Cost summary	Cost summary

Fittings

(7-) Fittings	(8-) Loose furniture equipment
(70) Vacant*	(80) Vacant*
Circulation	
(71) Circulation fittings	(81) Circulation loose furniture, equipment
Rest, work	
(72) Rest, work fittings	(82) Rest, work loose furniture, equipment
Culinary	
(73) Culinary fittings	(83) Culinary loose furniture, equipment
Sanitary	
(74) Sanitary, hygiene fittings	(84) Sanitary, hygiene loose furniture, equipment
Cleaning	
(75) Cleaning, maintenance fittings	(85) Cleaning, maintenance loose furniture, equipment
Storage	
(76) Storage, screening fittings	(86) Storage, screening loose furniture, equipment
(77) Special activity fittings	(87) Special activity loose furniture, equipment
(78) Other fittings	(88) Other loose furniture, equipment
(79) Parts, accessories, etc	(89) Parts, accessories etc
Cost summary	Cost summary

Other

(9-) External elements, other elements
(90) External works*
(91) Vacant*
(92) Vacant*
(93) Vacant*
(94) Vacant*
(95) Vacant*
(96) Vacant*
(97) Vacant*
(98) Other elements
(99) Parts, accessories etc
Cost summary

* See Appendix 6

(--) Sites, projects

(--) SITES, PROJECTS
Elements from two or more of element divisions (1-) to (8-) eg: structure plus fittings

Broader definition:
Sites, projects as a whole, ie sites for development, development of sites, building plus external works, building systems (several or all elements).
Building construction see (E4)
Structure if described separately see (2-)
External works if described separately see (90)

This position is used eg:
for library files for sites, building systems;
for location drawings, usually divided into eg:
block plans, general arrangement plans, elevations, sections etc for the whole project or for particular blocks, levels, rooms.

Some users of CI/SfB use (0) 'brackets nought' or (0-) instead of (--), so that the main classes of Table 1 can be coded (0) (1) (2) (3) (4) (5) (6) (7) (8) (9)
or
(0-) (1-) (2-) (3-) (4-) (5-) (6-) (7-) (8-) (9-)

If subdivision is needed it will often be enough to use only the headings below. Examples of their use are given:

(1-) Ground, substructure
Library files; dimensioned location drawings; assembly drawings; elemental specifications; bq sections

(2-) Structure
Library files; dimensioned location drawings

(3-) Secondary elements
Library files; project schedules of openings in structure; elemental specifications

(4-) Finishes
Library files; project schedules of finishes to structure; elemental specifications

(5-) Services (mainly piped)
Library files; location drawings showing position of pipe runs; elemental specifications

(6-) Services (mainly electrical)
Library files; location drawings; elemental specifications

(7-) Fittings
Library files; location drawings showing position of fittings; elemental specifications

(8-) Loose furniture, equipment
Library files; location drawings showing position of client's fittings

(9-) External works
Library files; location drawings; elemental specifications; bq sections

(1–) Ground, substructure

(1–) GROUND, SUBSTRUCTURE
Elements below underside of screed or, where no screed exists, to underside of lowest floor finish or pavement
Broader definition:
Ground, substructure as a whole, site (ie infrastructure)
External works if described separately see (90.1)/(90.4)

If subdivision of this class is needed it will often be enough to use only the headings below. Examples of their use are given:

(11) Ground
Library files
(13) Floor beds
Library files; assembly drawings, elemental specifications, bq sections
(16) Retaining walls, foundations
Separate library files for retaining walls, foundations; assembly drawings, elemental specifications, bq sections
(17) Pile foundations
Useful where piling needs to be shown separately from other types of foundations
(19) Minor parts of substructure
Library file
(19) Summary of this group
Cost plans

---

(10) Vacant

(11) GROUND

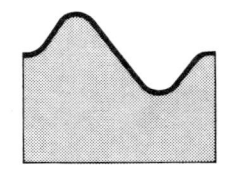

Ground for construction eg: excavated and filled ground, stabilised ground; underground
Broader definition:
Ground, earth shapes as a whole; site morphology, ground engineering
External works if described separately see (90.1)/(90.4)
If necessary, subdivide eg:
(11.1) Ground relief eg:
prominences, mounds, embankments; depressions, hollows, cuttings.
Ground inclination eg:
flat ground; slopes including sheer slopes, terraced slopes
(11.3) Ground composition eg:
stabilised ground
(11.4) Ground water
Ground underwater if described separately see (13.6)
(11.5) Underground eg:
tunnels, underground ducts including culverts; land drainage unless at (52)
(11.8) Other types of ground elements
(11.9) Parts, accessories etc special to ground elements may be included here if described separately from specific types above eg:
excavations, trenches, fill

(12) Vacant

(13) FLOOR BEDS

Floors in direct contact with the ground, with integral finish, or up to underside of screed, or, where no screed exists, to underside of lowest floor finish or pavement
Broader definition:
Floors in direct contact with the ground, as a whole, eg:
suspended ground floors, hard floorbeds, hard and soft surfaces
Raft foundations if described separately see (16.4)
Suspended floors if described separately see (23)
Secondary elements in floors if described separately see (33)
Floor finishes if described separately see (43)
External works if described separately see (90.4)
If necessary, subdivide eg:
(13.1) Floors on the ground, suspended ground floors, pavements (hard surfaces)
External hard surfaces if described separately see (90.41)
(13.4) Planted, unplanted beds (soft surfaces)
Plant display containers (fittings) if described separately see (71.2)
External planted surfaces if described separately see (90.42)
(13.6) Ground underwater eg:
pools
External water surfaces if described separately see (90.43)
Ground water as a whole see (11.4)
(13.8) Other types of floor beds
(13.9) Parts, accessories etc special to floor beds may be included here if described separately from specific types above eg:
damp proofing, insulation, reinforcement to floor beds in general

(14) Vacant

(15) Vacant

(1–) Ground, substructure

(16) RETAINING WALLS, FOUNDATIONS

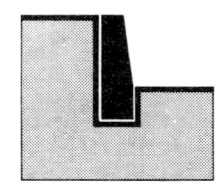

Retaining walls; foundations below lowest floor level damp proof course
Broader definition:
Retaining walls, foundations as a whole, including pile foundations
If necessary, subdivide eg:
- (16.2) Retaining walls eg:
  mass, gravity, crib retaining walls; flexible, cantilever retaining walls; counterfort retaining walls; buttressed retaining walls; tied retaining walls; diaphragm retaining walls; basement retaining walls (buildings); abutment retaining walls (bridges)
- (16.3) Water retaining elements eg:
  caissons
  Temporary cofferdams see (B2r)
- (16.4) Foundations eg:
  pad foundations, footings (rising walls); strip foundations; grillages; raft foundations
  Floor beds as a whole see (13)
- (16.8) Other types of retaining, foundation elements, apart from pile foundations
- (16.9) Parts accessories etc special to retaining, foundation elements may be included here if described separately from specific types above, eg:
  damp proofing, insulation, reinforcement to retaining walls, foundations in general

(17) PILE FOUNDATIONS

If described separately from class (16)

Pile foundations below lowest floor level damp proof course
Broader definition:
Pile foundations as a whole
External works if described separately see (90.1)
Temporary sheet piling see (B2r)
If necessary, subdivide eg:
- (17.1) Sheet piling
- (17.2) Replacement, in-situ formed pile foundations eg:
  bored, cased, uncased
- (17.3) Displacement, pre-formed pile foundations eg:
  driven, jacked, screwed
- (17.4) Small displacement
- (17.8) Other types of pile foundations
- (17.9) Parts accessories etc special to pile foundations may be included here if described separately from specific types above, eg:
  reinforcement to pile foundations in general

(18) OTHER SUBSTRUCTURE ELEMENTS

(19) PARTS, ACCESSORIES ETC SPECIAL TO SUBSTRUCTURE ELEMENTS
may be included here if described separately from specific types above eg:
damp proofing, insulation, reinforcement to substructure in general

Cost summary

(2–) Structure primary elements, carcass

(2–) STRUCTURE PRIMARY ELEMENTS, CARCASS
Structure above substructure
Structure carcass excluding secondary elements (3–); finishes (4–); services (5–) and (6–); fittings (7–) and (8–);
Broader definition:
Structure, carcass as a whole
Structure secondary elements if described separately see (3–)
Structure finishes if described separately see (4–)
Services if described separately see (5–)/(6–)
Fittings if described separately see (7–)/(8–)
External works if described separately see (90)

If subdivision of this class is needed it will often be enough to use only the headings below. Examples of their use are given:

(21) Walls, external walls
Library files; assembly, component drawings; elemental specifications, bq sections; cost plans

(22) Internal walls, partitions
Library files; assembly, component drawings; elemental specifications, bq sections; cost plans

(23) Floors
Library files; assembly, component drawings; elemental specifications, bq sections; cost plans

(24) Stairs
Library files; assembly, component drawings; elemental specifications, bq sections; cost plans

(27) Roofs
Library files; dimensioned location drawings; assembly, component drawings; elemental specifications, bq sections; cost plans

(28) Building frames
Library files; assembly, component drawings; elemental specifications, bq sections; cost plans

(29) Minor parts of structure
Library file

(29) Summary of this group
Cost plans

(2.1) Loadbearing wall structures eg:
box frame, cross wall structures
(2.2) Framed structures eg:
beam and column structures
Building frames see (28)
(2.3) Cellular structures eg:
slab and panel structures
(2.4) Stressed skin, geodesic structures eg:
folded plate structures
(2.5) Shell structures
(2.6) Tension structures eg:
cable structures, membrane structures
(2.8) Other types of structures eg:
air supported structures, cored structures (structures with structural core)
(2.9) Major divisions of structure, carcass eg:
external envelope, horizontal dividing elements
If necessary, subdivide eg:

(20) Vacant

(21) WALLS, EXTERNAL WALLS

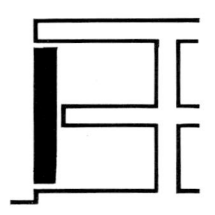

External walls carcass above lowest floor level damp proof course, ie excluding secondary elements, finishes, services and fittings unless described with carcass
Broader definition:
Walls as a whole, vertical dividing elements, both external and internal walls; external walls as a whole
Retaining walls as a whole see (16.2)
Internal walls if described separately see (22)
External walls secondary elements if described separately see (31)
Wall finishes externally if described separately see (41)
If necessary, subdivide eg:
(21.1) Loadbearing walls including cavity
(21.3) Non-loadbearing walls
(21.4) Curtain walls
(21.6) Framing and cladding walls
(21.8) Other types of walls eg:
compartment walls, gable walls, parapet walls
(21.9) Parts accessories etc special to walls, external walls may be included here if described separately from specific types above, eg:
damp proofing, insulation, reinforcement to external walls, walls in general

(22) INTERNAL WALLS, PARTITIONS

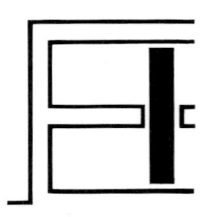

If described separately from class (21)

Internal walls carcass above lowest floor level damp proof course, ie excluding secondary elements, finishes, services and fittings unless described with carcass
Broader definition:
Internal walls, partitions as a whole:
Internal walls secondary elements, folding doors/partitions if described separately see (32)
Wall finishes internally if described separately see (42)
If necessary, subdivide eg:
(22.1) Loadbearing internal walls, cross walls including cavity
(22.3) Non-loadbearing internal walls eg:
fixed partitions, including post-to-panel, panel-to-panel; demountable partitions, including post-to-panel, panel-to-panel
(22.6) Framing and cladding, monolithic internal walls
(22.8) Other types of internal walls, eg:
party walls, separating walls, compartment walls
(22.9) Parts, accessories etc special to internal walls may be included here if described separately from specific types above, eg:
insulation, reinforcement to internal walls in general

If the distinction between **external** walls and **internal** walls is unimportant class (22) may be wholly or partially omitted.

(2–) Structure primary elements, carcass (2–)

(23) FLOORS, GALLERIES

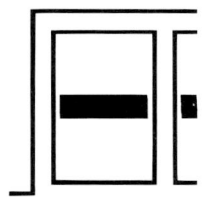

Suspended floors, galleries, balconies carcass, ie excluding secondary elements, finishes, services and fittings unless described with carcass
Broader definition:
Floors, suspended floors, galleries, balconies as a whole, including ceilings
Floor beds see (13)
Secondary floors if described separately see (33)
Suspended ceilings if described separately see (35)
Floor finishes if described separately see (43)
Platform fittings see (71.3)
If necessary, subdivide eg:
- (23.2) Monolithic, slab floors eg:
  one direction spanning slab floors;
  two direction spanning slab floors
- (23.4) Assembled, composite floors eg:
  beam and covering floors;
  joist plus infill floors;
  beam plus slab floors
- (23.7) Galleries, balconies
- (23.8) Other types of floors eg:
  party floors, separating floors, compartment floors
- (23.9) Parts, accessories etc special to floors, galleries may be included here if described separately from specific types above, eg:
  insulation, reinforcement to floors, galleries in general

(24) STAIRS, RAMPS

Stairs, ramps carcass ie excluding secondary elements, finishes, services and fittings unless described with carcass. Also lift wells carcass if shown separately
Broader definition:
Stairs, ramps, including soffits; vertical circulation elements as a whole
Stairs balustrades if described separately see (34)
Stairs finishes if described separately see (44)
If necessary, subdivide eg:
- (24.1) Straight stairs, including straightflight stairs
- (24.3) Dog leg stairs, other straight stairs
- (24.4) Curved stairs including helical stairs (spiral stairs)
- (24.5) Other types of stairs by amount of turn, open well stairs, escape stairs
- (24.6) Ladders, step irons, sliding poles
- (24.7) Ramps
- (24.8) Other types of vertical circulation elements eg:
  lift shaft
  Lifts, transport services see (66.1)
- (24.9) Parts, accessories etc special to stairs, ramps, vertical circulation elements may be included here if described separately from specific types above, eg:
  reinforcement to vertical circulation elements in general

(25) Vacant

(26) Vacant

(27) ROOFS

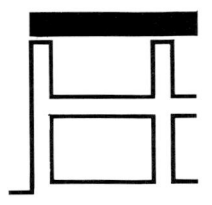

Roofs carcass excluding secondary elements, finishes, services, and fittings unless described with carcass
Broader definition:
Roofs as a whole, including ceilings
Suspended ceilings if described separately see (35)
Roof secondary elements if described separately see (37)
Roof finishes if described separately see (47)
If necessary, subdivide eg:
- (27.1) Flat roofs, terraces, platform roofs – up to 10° above horizontal
- (27.2) Pitched roofs eg:
  1 Single pitch, cut, lean-to, sloping
  2 Double pitch, saddleback
  3 Four pitch, hipped, helm
  8 Other types of pitched roofs eg: Mansard
- (27.4) Folded plate roofs
- (27.5) Other roofs by form eg:
  cylindrical roofs, conoïdal roofs, hyperbolic paraboloid roofs, domes, cupolas, steeples, spires
- (27.6) Roofs by structure eg:
  shell roofs, arched roofs, vaulted roofs, suspended roofs, space frame roofs
- (27.7) Cantilevered roofs, canopies
- (27.8) Other types of roofs eg:
  northlight roofs, gabled roofs, retractable roofs
- (27.9) Parts, accessories etc special to roofs may be included here if described separately from specific types above, eg:
  decking, trusses, damp proofing, insulation, reinforcement to roofs in general

(28) BUILDING FRAMES, OTHER PRIMARY ELEMENTS

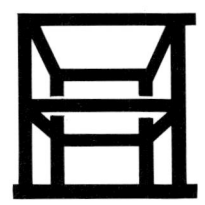

Building frame excluding secondary elements, finishes, services and fittings unless described with frame. Also chimneys, shafts
Broader definition:
Building frame, skeleton as a whole; other primary elements
If necessary, subdivide eg:
- (28.2) Column and beam frames, portal frames, column beam and slab frames
- (28.3) Column and slab frames
- (28.5) Column and cable 'frames'
- (28.6) Space frames as building frames
- (28.7) Other building frames eg:
  pin-jointed, rigid-jointed
- (28.8) Other types of primary elements eg:
  shafts, ducts, chimneys including flues
  Free standing chimneys see 161
  Flues described separately see (59)
- (28.9) Parts, accessories etc special to building frames etc may be included here if described separately from specific types above, eg:
  reinforcement to building frames in general

(29) PARTS, ACCESSORIES ETC SPECIAL TO PRIMARY ELEMENTS, CARCASS

may be included here if described separately from specific primary elements above eg:
damp proofing, insulation, reinforcement to primary elements in general

Cost summary

(3–) Secondary elements, completion of structure

(3–) SECONDARY ELEMENTS, COMPLETION OF STRUCTURE
Completions around and within openings in carcass, and other structure secondary elements
Structure as a whole see (2–)
Finishes if described separately see (4–)
External works if described separately see (90.2)/(90.3)

If subdivision of this class is needed it will often be enough to use only the headings below. Examples of their use are given:

(31) Secondary elements to walls, external walls
Library files for eg:
windows and external doors*; assembly, component drawings; elemental specifications, bq sections; cost plans
*Some users prefer to keep (31) for windows only and (32) for all doors; or to keep windows and all doors at (31) and omit (32).

(32) Secondary elements to internal walls, partitions
Library files for eg:
internal doors; assembly, component drawings; elemental specifications, bq sections; cost plans

(33) Secondary elements to floors
Library files for eg:
continuous access floors, trap doors; assembly, component drawings; elemental specifications, bq sections; cost plans, or include in (23)

(34) Secondary elements to stairs
Library files for eg:
staircase balustrades; assembly, component drawings; elemental specifications, bq sections; cost plans, or include in (24)

(35) Suspended ceilings
Library files for eg:
ceiling suspension systems; assembly, component drawings; elemental specifications, bq sections; cost plans, or include in (45)

(37) Secondary elements to roofs
Library files for eg:
eaves, rooflights; assembly, component drawings; elemental specifications, bq sections; cost plans, or include in (27)

(39) Minor parts of secondary elements
Library file for eg:
ironmongery

(39) Summary of this group
Cost plans

(30) Vacant

(31) SECONDARY ELEMENTS TO WALLS, EXTERNAL WALLS

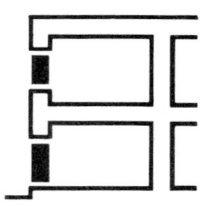

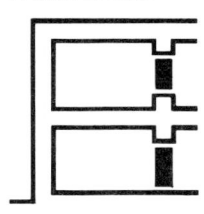

If described separately from class (21)

Completion around and within openings in external walls; and other external walls secondary elements eg:
porches
External walls as a whole see (21)
Internal wall secondary elements if described separately see (32)
Wall finishes externally if described separately see (41)
If necessary, subdivide eg:
(31.3) Window/door openings (including composite window and door openings), and parts to fill them such as entrance screens
(31.4) Window openings and windows to fill them
Internal window openings if described separately see (32.4)
(31.5) Doorways, entrances, exits and doors to fill them
Internal doorways if described separately see (32.5)
(31.7) Hatch openings and hatches to fill them
Internal hatches if described separately see (32.7)
(31.8) Other types of external wall, wall secondary elements eg:
barred openings, louvred openings
(31.9) Parts, accessories etc special to walls, external walls secondary elements may be included here if described separately from special types above, eg:
lintels, ironmongery to wall openings in general

The following subdivision of windows, doors, hatches etc can be applied to all (3–) classes:
1 Fixed
2 Hung
3 Pivoted, projected, revolving, louvres
4 Sliding, folding
5 Vertical sliding folding, roller shutters
6 Overhead opening, up and over
8 Other
9 Parts eg: ironmongery
For example:
(31.34) Sliding patio doors/windows
(31.44) Sliding windows
(31.54) Sliding doors

(32) SECONDARY ELEMENTS TO INTERNAL WALLS, PARTITIONS

If described separately from classes (22), (31)

Completion around and within openings in internal walls; and other internal walls secondary elements
Internal walls, partitions as a whole see (22)
Wall finishes internally if described separately see (42)
If necessary, subdivide eg:
(32.3) Internal window/door openings (including composite window and door openings), and parts to fill them
(32.4) Internal window openings and windows to fill them eg:
borrowed lights
Lay lights see (35.3)
(32.5) Internal doorways, room divider openings and parts to fill them
(32.7) Internal hatch openings, serveries and parts to fill them
(32.8) Other types of internal wall secondary elements eg:
barred openings
(32.9) Parts, accessories etc special to internal walls secondary elements may be included here if described separately from specific types above eg:
lintels, ironmongery to internal wall openings in general

If the distinction between external wall secondary elements and internal wall secondary elements is unimportant, class (32) may be omitted

(3–) Secondary elements, completion of structure (3–)

(33) SECONDARY ELEMENTS TO FLOORS

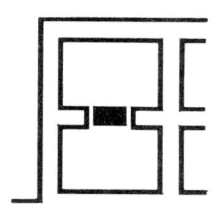

If described separately from classes (13), (23)

Completion around and within openings in suspended floors and floor beds; and other floors, galleries secondary elements
Floor beds see (13)
Floors, suspended floors as a whole see (23)
Suspended ceilings see (35)
Floor finishes if described separately see (43)
Platform fittings see (71.3)
If necessary, subdivide eg:
(33.1) Secondary suspended floors eg:
access floors including continuous access floors, cavity floors; stages
Floating floors (floor finishes) see (43)
(33.2) Secondary floor beds eg:
machine bases, hearths
(33.5) Floor (door) openings eg:
trap doorways, and parts to fill them
Pavement lights see (37.47)
(33.8) Other types of floor, secondary elements eg:
barred openings
(33.9) Parts, accessories etc special to floors, galleries secondary elements may be included here if described separately from specific types above, eg:
reinforcement, ironmongery to floor openings in general

(34) SECONDARY ELEMENTS TO STAIRS

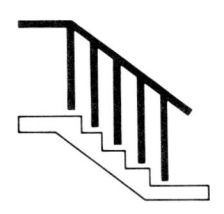

If described separately from class (24)

Completion around and within stair openings; and other stairs, ramps, lift shafts secondary elements eg:
stair balustrades, ramp balustrades, lift shaft guide rails
Stairs, ramps as a whole see (24)
Stair finishes if described separately see (44)

(35) SUSPENDED CEILINGS

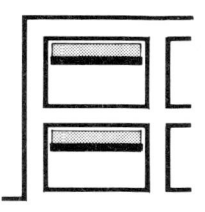

If described separately from classes (23), (27)

Completion under floors, roofs
Suspended floors as a whole see (23)
Roofs as a whole see (27)
Services if described separately see (5–), (6–)
If necessary, subdivide eg:
(35.1) Jointless suspended ceilings; panelled suspended ceilings constructed with panels, tiles, strips which may be solid, perforated, louvred, textured, sculptured etc
(35.2) Louvred suspended ceilings with separate vanes, not panels
(35.3) Ceiling openings and parts to fill them eg:
lay lights
(35.8) Other types of suspended ceiling elements eg:
ceiling walkways
(35.9) Parts, accessories etc special to suspended ceilings may be included here if described separately from specific types above eg:
insulation to suspended ceilings in general

(36) Vacant

(37) SECONDARY ELEMENTS TO ROOFS

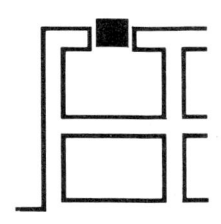

If described separately from class (27)

Completion around and within roof openings; and other roof secondary elements
Roofs as a whole see (27)
Suspended ceilings see (35)
Roof finishes if described separately see (47)
If necessary, subdivide eg:
(37.3) Roof window/door openings including composite window and door openings, and parts to fill them
(37.4) Roof window openings and parts to fill them eg:
1 Roof lights, including dome lights
2 Barrel lights, lantern lights, monitor lights, northlights
4 Dormer windows, eyebrow windows
5 Roof windows, skylights
7 Pavement lights
8 Other types of roof window openings
(37.5) Roof doorways eg:
trap doorways, access traps, and parts to fill them
(37.6) Roof eaves, parapets, balustrades if described separately
(37.8) Other types of roof secondary elements eg:
roof walkways
(37.9) Parts, accessories etc special to roof secondary elements may be included here if described separately from specific types above eg:
weather proofing, ironmongery to roof secondary elements in general

(38) OTHER SECONDARY ELEMENTS

eg: ducts, balustrades
Ducts as part of carcass see (28.8)

(39) PARTS, ACCESSORIES ETC SPECIAL TO SECONDARY ELEMENTS

may be included here if described separately from specific secondary elements above eg:
weather proofing, insulation, ironmongery to secondary elements in general

Cost summary

(4–) Finishes to structure

(4–) FINISHES TO STRUCTURE
Finishes applied to surface of structure including preparatory work, sub-layers or supports
Structure as a whole see (2–)
Completions, secondary elements see (3–)

If subdivision of this class is needed it will often be enough to use only the headings below. Examples of their use are given:

(41) Wall finishes, external
Library files; elemental specifications, bq sections; cost plans, or include in (21)

(42) Wall finishes, internal
Library files; elemental specifications, bq sections; cost plans

(43) Floor finishes
Library files; elemental specifications, bq sections; cost plans

(44) Stair finishes
Library files; elemental specifications, bq sections; cost plans, or include in (24)

(45) Ceiling finishes
Library files; elemental specifications, bq sections; cost plans

(47) Roof finishes
Library files; elemental specifications, bq sections; cost plans, or include in (27)

(49) Summary of this group
Cost plans

(40) Vacant

(41) WALL FINISHES, EXTERNAL

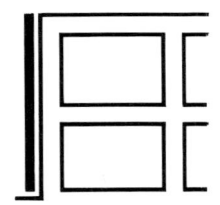

If described separately from classes (21), (31)

Finishes applied to external surface of external walls, including preparatory work, sub-layers or supports.
Parts, accessories etc special to wall finishes external may be included here
External walls as a whole see (21)
External walls secondary elements see (31)

(42) WALL FINISHES, INTERNAL

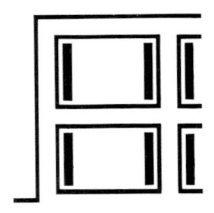

If described separately from classes (22), (32), (41)

Finishes applied to internal surface of internal and external walls, including preparatory work, sub-layers or supports
Parts, accessories etc special to wall finishes internal may be included here
Internal walls as a whole see (22)
Internal walls secondary elements see (32)

If the distinction between **external** wall finishes and **internal** wall finishes is unimportant, class (42) may be wholly or partially omitted

(4–) Finishes to structure

(43) FLOOR FINISHES

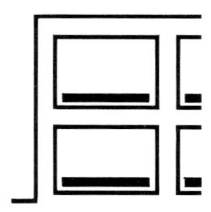

If described separately from classes (13), (23), (33)

Finishes applied to surface of suspended floors and floor beds, including preparatory work, sub-layers or supports. Includes floating floors (finishes)
Parts, accessories etc special to floor finishes may be included here, eg:
skirtings
Floor beds as a whole see (13)
Floors as a whole see (23)
Floors secondary elements see (33)
Stair finishes see (44)

(44) STAIR FINISHES

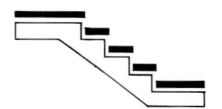

If described separately from (24), (34)

Finishes applied to surface of stairs, ramps, including preparatory work, sub-layers or supports.
Parts, accessories etc special to stair finishes may be included here eg:
stair nosings
Stairs as a whole see (24)
Stairs, ramps secondary elements see (34)

(45) CEILING FINISHES

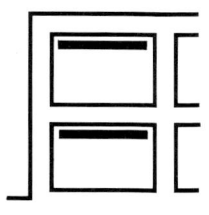

If described separately from (35)

Finishes applied to soffits of floors, roofs, to suspended ceilings, to other overhead surfaces, including preparatory work, sublayers or supports
Parts, accessories etc special to ceiling finishes may be included here, eg:
ceiling coves, ceiling cornices
Floors as a whole see (23)
Roofs as a whole see (27)
Suspended ceilings as a whole see (35)

(46) Vacant

(47) ROOF FINISHES

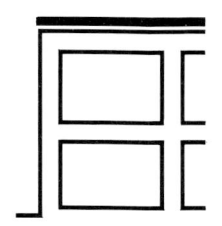

If described separately from (27), (37)

Finishes applied to surface of roof, including preparatory work, sub-layers or supports.
Parts, accessories etc special to roof finishes may be included here, eg:
roof edgings, fascias, flashings
Roofs as a whole see (27)
Roof secondary elements see (37)

(48) OTHER FINISHES TO STRUCTURE
eg: decorations

(49) PARTS, ACCESSORIES ETC SPECIAL TO FINISHES TO STRUCTURE ELEMENTS
may be included here if described separately from specific finishes above, eg:
sub-layers or supports, flashings, edgings to structure finishes in general

Cost summary

(5–)

(5–) SERVICES
Mainly piped and ducted services excluding electrical inputs and fittings connected to two or more services
Broader definition:
Services as a whole
Multi-service fittings see (7–)
Services in external works if described separately see (90.5)
Temporary services see (B2c)

If subdivision of this class is needed it will often be enough to use only the headings below. Examples of their use are given:
(52) Waste disposal, drainage
Separate library files for waste disposal*, drainage; assembly, component drawings; elemental specifications, bq sections; cost plans
*Some users keep waste disposal at (51)
(53) Hot and cold water supply
Library files; assembly, component drawings; elemental specifications, bq sections; cost plans
(54) Gases supply
Library files; assembly, component drawings; elemental specifications, bq sections; cost plans
(55) Refrigeration
Useful where refrigeration needs to be shown separately from air conditioning at (57)
(56) Space heating
Library files; assembly, component drawings; elemental specifications, bq sections; cost plans
(57) Air conditioning, ventilation
Library files; assembly, component drawings; elemental specifications, bq sections; cost plans
(59) Minor parts of piped services
Library files
(59) Summary of this group
Cost plans

(5.1) Power sources for services
  1 Solid fuel
  2 Oil
  3 Gas
  4 Electricity
  5 Solar radiation
  6 Other eg: occupants, machines, waste matter
  This subdivision can be applied to (5–) classes as appropriate
  See eg note at (56.1)
(5.2) Integrated services eg:
  heart units
(5.7) District or community, centralised, localised services

(5–) Services

(50) Vacant

(51) Vacant

(52) WASTE DISPOSAL, DRAINAGE

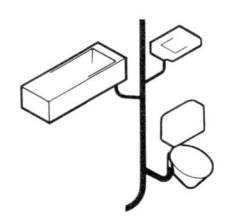

Waste disposal, drainage services excluding installations described with other elements, eg:
sink waste
Broader definition:
Disposal services as a whole
If necessary, subdivide eg:
(52.1) Refuse, rubbish, garbage disposal
  3 Dry carriage eg: chute and hopper systems, vacuum systems
  4 Manual including dustbins, sacks
  8 Other types of refuse disposal including local disposal eg: incinerators, crushers, grinders, shredders, compactors, balers
  Special disposal fittings if described separately see (73), (74), (75)
(52.2) Gaseous waste
(52.3) Sewage disposal
  Foul drainage eg:
  soil, foul water waste including waterborne refuse
(52.4) Petrol, chemical wastes eg:
  organic solvents, corrosive, radio-active effluents
(52.5) Natural water drainage eg:
  rainwater; surface water
  Land drainage see (11.5)
(52.6) Internal drainage (above ground drainage)
(52.7) Below ground drainage unless at (90.5), including local storage and treatment eg:
  septic tanks, cesspools
(52.8) Other types of waste disposal, drainage services
(52.9) Parts, accessories etc special to waste disposal, drainage services may be included here if described separately from specific types above eg:
  traps, ventilation to waste disposal, drainage services in general

(53) LIQUIDS SUPPLY

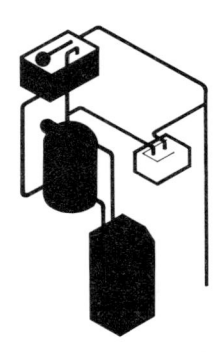

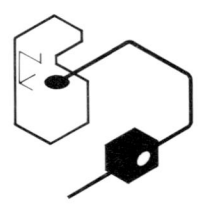

Liquids supply services, excluding installations described with other elements eg:
Header tanks for central heating services see (56.9)
Broader definition:
Liquids supply services as a whole
If necessary, subdivide eg:
(53.1) Cold water eg:
  potable (drinking) water; chilled water
(53.3) Hot water eg:
  Hot water from common supply
(53.5) Hot water from individual appliances
(53.6) Other water supply services eg:
  filtration and treatment of water for special purposes; softened water, de-ionised water
(53.7) Petrol, oil
(53.8) Other types of liquids supply services
  Soap dispensers see (74.7)
(53.9) Parts, accessories etc special to liquids supply services may be included here if described separately from specific types above, eg:
  cisterns, ball valves, lagging to liquids supply services in general

(54) GASES SUPPLY

Gases supply services, excluding installations described with other elements eg:
Gas heaters for hot water supply see (53.5)
Broader definition:
Gases supply services as a whole
If necessary, subdivide eg:
(54.1) Fuel gas, combustible gas supply eg:
  natural gas; propane, butane

(5–) Services (5–)

- **(54.2)** Vapour supply eg:
  steam supply and condensate services including vacuum, atmospheric and pressurised services, heated steam services
- **(54.3)** Air supply eg:
  compressed air
  Warm air heating services see (56.5)
  Ventilation, air conditioning services see (57)
- **(54.4)** Other gas supply eg:
  medical gases, industrial gases
- **(54.5)** Vacuum supply
  Vacuum refuse removal see (52.13)
- **(54.8)** Other types of gases supply services eg:
  supply of solids using gas pressure or vacuum
  Pneumatic message systems see (64.8)
- **(54.9)** Parts, accessories etc special to gases supply may be included here if described separately from specific types above, eg:
  meters for gases supply services in general

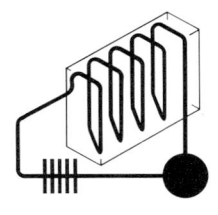

(55) SPACE COOLING

Space cooling, refrigeration services excluding installations described with other elements eg:
Culinary refrigerators see (73.5)
Broader definition:
Space cooling, refrigeration services as a whole
If necessary, subdivide eg:
- **(55.1)** Central refrigeration
- **(55.5)** Local refrigeration
  Culinary refrigerators, freezers see (73.5)
- **(55.8)** Other types of space cooling services
- **(55.9)** Parts, accessories etc special to space cooling, refrigeration services may be included here if described separately from specific types above, eg:
  lagging for space cooling, refrigeration services in general

(56) SPACE HEATING

Space heating services excluding installations described with other elements eg:
Space heating for air conditioning services see (57)
Broader definition:
Space heating services as a whole
If necessary, subdivide eg:
- **(56.1)** Heating by power source
  Subdivide here and in other subclasses below as at (5.1) eg:
  solid fuel heating (56.11)
- **(56.2)** Communal heating (remote centre with mains and local distribution) eg:
  district heating (supplied to any consumer within economic distance of heat centre);
  group heating (supplied to groups of buildings in common ownership)
- **(56.3)** Central heating (local centre and/or distribution) eg:
- **(56.4)** Hot water, steam distribution
- **(56.5)** Warm air distribution
- **(56.6)** Electrical distribution eg:
  coils, storage heaters
- **(56.7)** Other types of central heating
- **(56.8)** Other types of space heating services eg:
  direct space heating (local appliances not part of a central system)
- **(56.9)** Parts, accessories etc special to space heating services may be included here if described separately from specific types above eg:
  lagging for space heating services in general

(57) AIR CONDITIONING, VENTILATION

Air conditioning services (with air treatment) or ventilation services (without air treatment) excluding installations described with other elements
Broader definition:
Air conditioning, ventilation services as a whole
If necessary, subdivide eg:
- **(57.1)** Central air conditioning eg:
  air heating and cooling
- **(57.2)** Air heating only
- **(57.3)** Local air conditioning (local appliances not part of a central system) eg:
  air heating and cooling
- **(57.4)** Air heating only
- **(57.5)** Air treatment eg:
  heating, cooling, humidifying, drying, filtration, pressurisation, recirculation, extraction
- **(57.6)** Mechanical ventilation services (no air treatment), including powered supply or extract, or both eg:
  central ventilation
- **(57.7)** Unit ventilation (local appliances not part of a central system)
- **(57.8)** Other types of air conditioning, ventilation services
- **(57.9)** Parts, accessories etc special to air conditioning, ventilation services may be included here if shown separately from specific types above eg:
  lagging to air conditioning, ventilation services in general

(58) OTHER PIPED, DUCTED SERVICES

(59) PARTS, ACCESSORIES ETC SPECIAL TO PIPED, DUCTED SERVICES ELEMENTS

may be included here if described separately from specific services above eg:
flues, boilers, lagging to piped, ducted services in general
Flues as part of carcass see (28.8)

Cost summary

(6–) Services, mainly electrical

(6–) SERVICES, MAINLY ELECTRICAL

Mainly electrical services excluding fittings connected to two or more services, installations described with other elements eg: electrical circuits for space heating at (56)
Broader definition:
Electrical services as a whole
Multi-service fittings see (7–)
Electrical services in external works if described separately see (90.6)

If subdivision of this class is needed it will often be enough to use only the headings below. Examples of their use are given:

(61) Electrical supply
Library files; location, assembly, component drawings; elemental specifications, bq sections; cost plans

(62) Power
Useful where electrical power supply needs to be shown separately from electrical supply (source and mains) at (61).

(63) Lighting
Useful where lighting needs to be shown separately from electrical supply (source and mains) at (61).

(64) Communications
Library files; assembly, component drawings; elemental specifications, bq sections; cost plans

(66) Transport
Library files; location, assembly, component drawings; elemental specifications, bq sections; cost plans

(68) Security
Library files; assembly, component drawings; elemental specifications, bq sections; cost plans

(69) Minor parts of electrical services
Library files

(69) Summary of this group
Cost plans

---

(60) Vacant

(61) ELECTRICAL SUPPLY

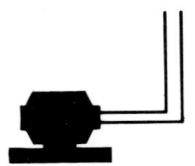

Electrical load centre and mains including mains intake, control gear, distribution to local subcircuit boards or to equipment permanently attached to electrical installation; excluding installations described with other elements
Broader definition:
Electrical supply services as a whole
If necessary, subdivide eg:
- (61.1) Radial distribution: mains intake to local control gear
- (61.2) Ring main distribution: mains intake to local control gear
- (61.3) Rising main distribution: mains intake to local control gear
- (61.6) Public mains supply eg:
  high voltage supply $> 650v$;
  medium voltage supply $250v–650v$;
  low voltage supply $\not> 250v$
- (61.7) Privately generated supply eg:
  emergency/standby supply
- (61.8) Other types of electrical supply services
- (61.9) Parts, accessories etc special to electrical supply services may be included here if described separately from specific types above eg:
  distribution boards, conduit, meters for electrical supply services in general

(62) POWER

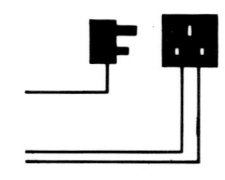

If described separately from (61)

Electric power subcircuits from local distribution boards to general purpose socket outlets; excluding installations described with other elements
Parts, accessories etc special to power distribution may be included here eg:
socket outlets
Broader definition:
Electrical power supply as a whole
Electrical heating see (56)

(63) LIGHTING

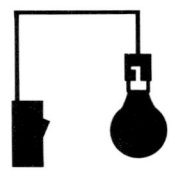

If described separately from (61)

Electric lighting circuits from local distribution boards, including lighting fittings; excluding installations described with other elements eg:
Illuminated signs see (71.1)
Broader definition:
Electric lighting supply as a whole
If necessary, subdivide eg:
- (63.1) General lighting, localised lighting
- (63.2) Local lighting including spot lighting, other display lighting
- (63.6) Floodlighting
- (63.8) Other types of lighting services eg:
  emergency lighting; flameproof, water proof lighting; gas lighting
- (63.9) Parts, accessories etc special to lighting may be included here if described separately from specific types above eg:
  lighting fittings eg:
  trough, bowl, louvred, recessed, flush, pendant, portable
  lighting sources eg: filament (incandescent) including tungsten; discharge including fluorescent; other lighting by lamp type

(6–) Services, mainly electrical (6–)

(64) COMMUNICATIONS

Communications services, including those linked to public networks, and complete private networks; excluding installations described with other elements eg:
Audio-visual security services see (68.2)
Broader definition:
Communications services as a whole
If necessary, subdivide eg:
(64.1) Visual, including audio-visual
  1 Television
  2 Film projection
  3 Indicating eg: staff location
  8 Other types of visual, audio visual services
(64.3) Audio
  1 Radio
  2 Telephone and intercom including telephone booths
  3 Relaying eg: public address
  4 Recording eg: central dictation
  5 Indicating eg: staff location
  8 Other types of audio services
(64.4) Signals other than visual or audio
  1 Telegraph
  2 Teleprinter, telex
  3 Facsimile transmission
  4 Data transmission
  8 Other types of signals
(64.5) Synchronous clocks
(64.8) Other types of communications services eg:
  pneumatic message systems
(64.9) Parts, accessories etc special to communication services may be included here if described separately from specific types above eg:
  loudspeakers, exchanges, handsets, conduit for communications services in general

(65) Vacant

(66) TRANSPORT

Transport, mechanical circulation services excluding installations described with other elements
Broader definition:
Transport, mechanical circulation services as a whole
Transport plant see (B3)
If necessary, subdivide eg:
(66.1) Lifts eg:
  passenger, goods, service lifts; paternosters, telescopic shaft lifts; scissor lifts; hydraulic lifts
(66.2) Other types of internal lifts, hoists
  Lift shafts if described separately see (24.8)
(66.3) Travelling cradles etc eg:
  for facade cleaning
(66.4) Escalators
(66.5) Conveyors eg:
  moving pavements; turntables; pneumatic, gravity conveyors
(66.7) Cranes eg:
  bridge type, gantry cranes
  Cranes as construction plant see (B3q)/(B3v)
(66.8) Other types of transport services eg:
  mechanised transport/storage services eg: palletized systems
(66.9) Parts, accessories etc special to transport services may be included here if described separately from specific types above eg:
  buzzers, motors, pulleys for transport services in general

(67) Vacant

(68) SECURITY, CONTROL, OTHER SERVICES

Security, protection, control services excluding installations described with other elements
Broader definition:
Protection, control services as a whole
Protection plant see (B1)
If necessary, subdivide eg:
(68.1) Security/fire protection services
(68.2) Security services eg:
  intruder, burglar detection, surveillance, alarm
(68.5) Fire protection services eg:
  2 Fire detection, alarm
  3 Fire fighting services
  4 Automatic eg: water (sprinklers), foam, carbon dioxide, dry chemical, vaporising liquid
  5 Manual eg: hose reels
  6 Portable eg: fire extinguishers, blankets
  8 Other types of fire protection services
(68.6) Other security, protection services eg:
  flood, lightning, bird nuisance
(68.7) Control services including process control, monitoring services eg:
  pneumatic, hydraulic control; electric, electronic, radio control; mechanical, clockwork control
(68.8) Other types of security, control services eg:
  sound control and attenuation
(68.9) Parts, accessories etc special to security, control services may be included here if described separately from specific types above eg:
  buzzers, bells, conduit for security, control services in general

(69) PARTS, ACCESSORIES ETC SPECIAL TO ELECTRICAL SERVICES ELEMENTS
may be included here if described separately from specific services above eg:
buzzers, conduit, outlets for electrical services in general

Cost summary

(7–) Fittings

(7–) FITTINGS
Fittings excluding loose furniture and equipment and installations described with other elements eg:
taps as part of water supply see (53.9)
Broader definition:
Fittings as a whole, including built-in or otherwise fixed fittings and loose furniture, equipment
Loose furniture, equipment if described separately see (8–)
Fittings in external works if described separately see (90.7)

If subdivision of this class is needed it will often be enough to use only the headings below. Examples of their use are given:

(71) Circulation fittings
Library files; assembly, component drawings; elemental specifications, bq sections
(72) Rest, work fittings
Library files; assembly, component drawings; elemental specifications, bq sections
(73) Culinary fittings
Library files; assembly, component drawings; elemental specifications, bq sections
(74) Sanitary, hygiene fittings
Library files; assembly, component drawings; elemental specifications, bq sections; cost plans
(75) Cleaning, maintenance fittings
Library files; assembly, component drawings; elemental specifications, bq sections
(76) Storage, screening fittings
Library files; assembly, component drawings; elemental specifications, bq sections
(77) Special activity fittings
Library files; assembly, component drawings; elemental specifications, bq sections
(79) Minor parts of fittings
Library files
(79) Summary of this group
Cost plans

(70) Vacant

(71) CIRCULATION FITTINGS

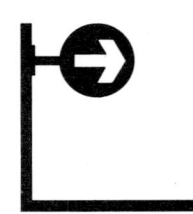

Built-in or otherwise fixed circulation fittings excluding installations described with other elements
These are fittings unconnected with any user activity at (72)/(77)
Broader definition:
Circulation fittings and furniture as a whole, including built-in or otherwise fixed fittings, and loose furniture, equipment
Loose furniture, equipment if described separately see (81)
If necessary, subdivide eg:
(71.1) Signs, symbols eg:
flagpoles, illuminated signs, emergency signs; lettering; notice boards
(71.2) Display fittings eg:
plant display containers
(71.3) Access fittings eg:
entrance mats; platforms, sets of steps
(71.8) Other types of circulation fittings
(71.9) Parts, accessories etc special to circulation fittings may be included here if described separately from specific types above

(72) REST, WORK FITTINGS

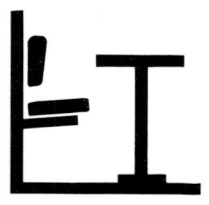

Built-in or otherwise fixed rest, work fittings, excluding installations described with other elements eg:
Culinary fittings see (73)
Broader definition:
Rest, work, fittings and furniture as a whole, including built-in or otherwise fixed fittings and loose furniture, equipment
Loose furniture, equipment if described separately see (82)
If necessary, subdivide eg:
(72.1) Rest fittings
Fittings for sleep eg:
suites including wardrobes etc; beds, bunks, cots, cradles
(72.2) Fittings for relaxation eg:
easy chairs, armchairs, couches
(72.3) Work fittings eg:
work stations, work benches, desks
(72.6) Benches, tables; seating, chairs
Dining tables, seating etc if described separately see (73.7), (83)
(72.8) Other types of rest, work fittings
(72.9) Parts, accessories etc special to rest, work fittings may be included here if described separately from specific types above
Upholstery see (78.3)

(73) CULINARY FITTINGS

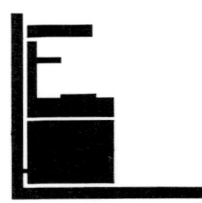

Built-in or otherwise fixed culinary fittings including catering fittings, but excluding installations described with other elements
Broader definition:
Culinary fittings as a whole, including built-in or otherwise fixed fittings and loose furniture, equipment
Culinary loose furniture, equipment if described separately see (83)
If necessary, subdivide eg:
(73.1) Culinary work fittings, eg:
work tops, suites
(73.2) Culinary sink units, draining boards, disposal units, washing-up machines
(73.4) Culinary processing, cooking fittings eg:
potato peelers; cookers, hotplates, boiling pans, fryers, grills, ovens; hoods
(73.5) Culinary storage fittings eg:
refrigerators, freezers; bains marie, hot cupboards, hot/cold storage/display fittings; larder fittings
(73.7) Bar counters, food counters, dining tables, seating etc
Loose fittings if described separately see (83)
(73.8) Other types of culinary, catering fittings eg:
food, drink vending machines; drinking fountains, water coolers
Table ware if described separately see (83)
(73.9) Parts, accessories etc special to culinary, catering fittings may be included here if described separately from specific types above

(7–) Fittings

(74) SANITARY, HYGIENE FITTINGS

Built-in or otherwise fixed sanitary, hygiene fittings, excluding installations described with other elements
Broader definition:
Sanitary, hygiene fittings as a whole, built-in or otherwise fixed fittings and loose furniture, equipment
Sanitary, hygiene loose furniture, equipment if described separately see (84)
If necessary, subdivide eg:
- (74.1) Sanitary suites
- (74.2) Washing fittings eg:
  - 2 Baths
  - 3 Washbasins, troughs, fountains
  - 4 Bidets, footbaths
  - 6 Shower fittings
  - 7 Sauna fittings
  - 8 Other types of washing fittings
- (74.3) Drying fittings eg:
  - towel cabinets; hand driers
- (74.4) Disposal fittings eg:
  - 4 Closets, including WC's, chemical, earth
  - 5 Urinals, including stall and bowl
  - 8 Other sanitary disposal fittings eg: macerating, packaging, incinerating, chemical disposal fittings
- (74.7) Supply fittings eg:
  - soap dispensers, sanitary goods dispensers
- (74.8) Other types of sanitary, hygiene fittings eg:
  - vanitory fittings, mirrors
- (74.9) Parts, accessories etc special to sanitary, hygiene fittings may be included here if described separately from specific types above

(75) CLEANING, MAINTENANCE FITTINGS

Built-in or otherwise fixed cleaning, maintenance fittings excluding installations described with other elements eg:
Culinary sinks see (73.2)
Broader definition:
Cleaning, maintenance fittings as a whole including built-in or otherwise fixed fittings, and loose furniture, equipment
Cleaning, maintenance loose equipment if described separately see (85)
If necessary, subdivide eg:
- (75.1) Washing fittings eg:
  - washing machines, sinks
- (75.3) Drying fittings eg:
  - airing fittings, drying machines
- (75.4) Pressing, ironing fittings
- (75.8) Other types of cleaning, maintenance fittings eg:
  - dry cleaning fittings
  - *Vacuum cleaners, floor polishers if described separately see (85)*
- (75.9) Parts, accessories etc special to cleaning, maintenance fittings may be included here if described separately from specific types above

(76) STORAGE, SCREENING FITTINGS

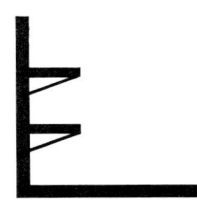

Built-in or otherwise fixed storage, screening fittings, excluding installations described with other elements eg:
Culinary storage see (73.5)
Broader definition:
Storage, screening fittings as a whole, including built-in or otherwise fixed fittings and loose furniture, equipment
Storage loose furniture, equipment if described separately see (86)
If necessary, subdivide eg:
- (76.1) Composite storage fittings, different configurations of cupboards, drawers, shelves eg:
  - storage walls
- (76.2) Cupboards fittings
  - which may contain shelves, drawers, suspended storage
- (76.3) Drawers fittings
  - which may contain suspended storage
- (76.4) Shelving, racking fittings
  - which may contain cupboards, drawers, suspended storage
- (76.5) Suspended storage fittings
- (76.6) Storage fittings with additional facility eg:
  - Secure storage fittings including safes, lockers, hot/cold storage
  - *Strong boxes, travel equipment if described separately (86)*
  - Other types of storage fittings eg: storage by what stored, such as clothes storage
  - *Food storage see (73.5)*
- (76.7) Screening fittings eg:
  - screens, blind boxes, blinds; shutters; curtain tracks, curtains, drapes
- (76.8) Other types of storage, screening fittings
- (76.9) Parts, accessories etc special to storage, screening fittings may be included here if described separately from specific types above.

(77) SPECIAL ACTIVITY FITTINGS

Built-in or otherwise fixed special activity fittings excluding installations described with other elements
Broader definition:
Special activity fittings as a whole, including built-in or otherwise fixed fittings and loose furniture, equipment
This class can be used for Table 0 facilities when they occur in the form of fittings and have not been listed so far in this element division
They can be coded eg:
562(77) Trampoline, filed at class 562; o~~r~~
(77)562 Trampoline, filed at this position

(78) OTHER FITTINGS

Other built-in or otherwise fixed fittings excluding installations described with other elements
Broader definition:
Other fittings as a whole, including built-in or otherwise fixed fittings and loose furniture, equipment
If necessary, subdivide eg:
- (78.3) Soft furnishings including upholstery
- (78.6) Works of art

(79) PARTS, ACCESSORIES ETC SPECIAL TO FITTINGS ELEMENTS

may be included here if described separately from the specific types above eg:
castors

Cost summary

(8–)   **(8–)**   **Loose furniture, equipment**

(8–) LOOSE FURNITURE, EQUIPMENT
Loose fittings, furniture, equipment

> A library file may eventually be required for each of the headings below but the (8–) group of headings as a whole is unlikely to be used for project information unless for the positioning of 'client's fittings'.

(80) Vacant

(81) CIRCULATION LOOSE FURNITURE, EQUIPMENT

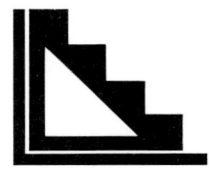

If described separately from class (71)

(82) REST, WORK LOOSE FURNITURE, EQUIPMENT

If described separately from class (72)

(83) CULINARY LOOSE FURNITURE, EQUIPMENT

If described separately from class (73)

(84) SANITARY, HYGIENE LOOSE FURNITURE, EQUIPMENT

If described separately from class (74)

(85) CLEANING, MAINTENANCE, LOOSE FURNITURE, EQUIPMENT

If described separately from class (75)

(86) STORAGE, SCREENING LOOSE FURNITURE, EQUIPMENT

If described separately from class (76)

(87) SPECIAL ACTIVITY LOOSE FURNITURE, EQUIPMENT
If described separately from class (77)

(88) OTHER LOOSE FURNITURE, EQUIPMENT
If described separately from class (78)

(89) PARTS, ACCESSORIES ETC SPECIAL TO LOOSE FURNITURE, EQUIPMENT

Cost summary

If the distinction between **fixed** and **loose** fittings, furniture and equipment is unimportant, class (8–) may be wholly or partially omitted

(9-) External elements, other elements (9-)

(9-) EXTERNAL ELEMENTS, OTHER ELEMENTS

(90) External works
Library files; location drawings; assembly, component drawings; elemental specifications, bq sections; cost plans

(99) Minor parts common to several element divisions
Library files for eg: damp proof courses, reinforcement, insulation

(99) Summary of all groups
Cost plans

---

(90) EXTERNAL WORKS
Work outside the external face of the external wall where shown separately from building works

(90.1) Ground preparation in external works if not at (1-) eg:
ground clearing, shaping

(90.2) Minor structures in external works if not at (1-)/(8-) eg:
ancillary shelters, sheds etc

(90.3) Enclosures in external works if not at (1-)/(3-) eg:
site walls, fences, trellis, gates, barriers, bollards

(90.4) Ground surface treatments in external works if not at (1-)
1 Hard surfaces eg:
  vehicular hardstandings, roads, paths, steps;
  edgings, trim eg:
  tree grilles, cattle grids
2 Soft surfaces, planted surfaces
3 Water surfaces, pools
8 Other types of ground surface treatments

(90.5) Piped services in external works if not at (5-)
1 Drainage eg:
  septic tanks
2 Other types of piped services eg:
  heating mains, pumps, jets, soil heating

(90.6) Electrical services in external works if not at (6-) eg:
outdoor lighting

(90.7) Fittings in external works if not at (7-) eg:
hoardings, outdoor benches, outdoor plant containers etc, outdoor play, sports equipment (as parts of project)

(90.8) Special landscaping in external works eg:
tennis courts, swimming pools (as part of project) if not divided between classes above

Two digits may be preferred instead of the codes above eg:
(91) rather than (90.1)

---

(98) OTHER ELEMENTS

(99) PARTS, ACCESSORIES ETC COMMON TO TWO OR MORE ELEMENT DIVISIONS (1-)/(7-)
may be included here if described separately from specific elements above

Cost summary

At class (99), as throughout Table 1, Appendix 1 provides a subdivision if the amount of material justifies this eg:
(99.53) Damp proof courses, tanking, insulation;
(99.71) Frames in general;
(99.79) Reinforcement.

All the subjects in Appendix 1 can be classified at appropriate levels in Table 1 if placed at classes ending in 9 as well as at the specific Table 1 classes eg:
(31.49) Window frames
(31.59) Door frames
(31.9) Window and door frames

The SfB Agency UK will not use Appendix 1 in classifying literature for manufacturers.

Table 2

Constructions, forms

Key

Scope
Parts of particular forms which combined to make the elements in Table 1. Each is characterized by the main product of which it is made, eg: Blockwork is characterized by blocks. These products are also included in the table.

Structure
Two main divisions:
Block and section construction forms coded F to I;
Sheet forms coded K to V.
Joints at Z are common to the whole of Tables 1 and 2, and come at the end of Table 2.

Changes
Most changes from the last edition are at a detailed level. They make the table a useful tool for the classification of information on constructions such as 'Brickwork' as well as products such as 'Bricks'.

Coding
Coding is capital letters. 'O' is omitted because of possible confusion with 'nought'.
Code positions marked 'vacant' may be used for any purpose in private application but not on published literature.

Use
Where two classes seem equally relevant – this is a matter for personal judgement – use the first in the absence of any private rule to the contrary. When using the table for a particular purpose, consider:

Simplicity of use
Use codes only if they serve a useful purpose.

Compound subjects
Adopt rules for dealing consistently with compound subjects which relate to two or more different constructions or products.

Matrix

A	Constructions, forms		**Sheets K/V**	W	{ Planting work / Plants
B	Vacant	K	{ Quilt work / Quilts	X	{ Work with components / Components
C	Excavation and loose fill work	L	{ Flexible sheet work (proofing) / Flexible sheets (proofing)	Y	{ Formless work / Products
D	Vacant	M	{ Malleable sheet work / Malleable sheets	Z	Joints*
E	Cast in situ work	N	{ Rigid sheet overlap work / Rigid sheet for overlapping		
	Blocks F/G	P	Thick coating work		
F	{ Blockwork, brickwork / Blocks, bricks	Q	Vacant		
G	{ Large block, panel work* / Large blocks, panels	R	{ Rigid sheet work / Rigid sheets		
	Sections H/I	S	{ Rigid tile work / Rigid tiles		
H	{ Section work / Sections	T	{ Flexible sheet and tile work / Flexible sheets, tiles		
I	{ Pipework / Pipes	U	Vacant		
J	{ Wire work, mesh work / Wires, meshes	V	Film coating and impregnation work		

*See Appendix 6

A/G

A Constructions, forms
C Excavation and loose fill work
E Cast in situ work
F Blockwork, brickwork
G Large block, panel work

Examples of the uses of the codes and headings are:

A Preliminaries and general conditions
Specifications, bq sections
(library files see (A) Table 4)

B Demolition and shoring work
Specifications, bq sections
(library files see (B) Table 4)

C Excavation and loose fill work
Library files; specifications, bq sections

E Cast in-situ work
Library files; specifications, bq sections

F Blockwork, brickwork
Library files; specifications, bq sections

G Large block, panel work
Library files; specifications, bq sections

H Section work
Library files; specifications, bq sections

I Pipe work
Library files; specifications, bq sections

J Wire work, mesh work
Library files; specifications, bq sections

K Quilt work
Library files; specifications, bq sections

L Flexible sheet work (proofing)
Library files; specifications, bq sections

M Malleable sheet work
Library files; specifications, bq sections

N Rigid sheet overlap work
Library files; specifications, bq sections

P Thick coating work
Library files; specifications, bq sections

R Rigid sheet work
Library files; specifications, bq sections

S Rigid tile work
Library files; specifications, bq sections

T Flexible sheet work
Library files; specifications, bq sections

V Film coating and impregnation work
Library files; specifications, bq sections

W Planting work
Library files for eg: nursery stock; specifications, bq sections

X Work with components
Library files for eg: fixings; specifications, bq sections

Y Formless work
Library files for eg: making mortar; specifications, bq sections

Z Joints
Library files

A CONSTRUCTIONS, FORMS
Constructions, product forms as a whole, of two or more of specific classes B to X eg:
pipe work plus film coating work, blocks plus rigid sheets

Used in specification applications for
PRELIMINARIES AND GENERAL CONDITIONS
Building construction see (E4)

B Vacant

Used in specification applications for
DEMOLITION AND SHORING WORK
including underpinning
Substructure see (1–)
Temporary shores see (B2d)

C EXCAVATION AND LOOSE FILL WORK

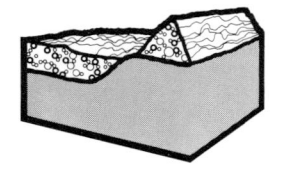

Work with excavated materials and loose fill
Types eg:
Excavating work, loose filling work, loose paving work
Accessories

D Vacant

E CAST IN SITU WORK

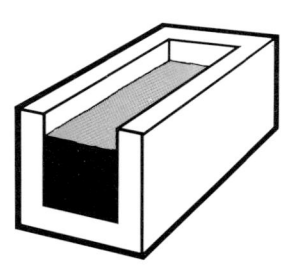

Casting work both on and off site
Types eg:
Unreinforced (plain) casting work, precasting work, casting in situ work, battery casting work, reinforced casting work, prestressed casting work
Accessories

F BLOCKWORK, BRICKWORK

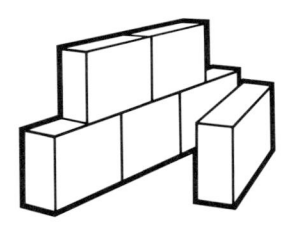

Work with blocks
Parquet tiling work see S
Types eg:
Blockwork eg: block walling; block paving; fair faced, flush, indented and projecting blockwork; uncoursed and brought to courses blockwork; raking bond, basket weave bond, stack bond, stretcher bond, header bond, stretcher/header bond and honeycomb bond blockwork etc

Brickwork eg: brick walling; brick paving; fairfaced, flush, indented and projecting brickwork; half, one, one and a half, two brick brickwork etc; raking bond, basket weave bond, stack bond, stretcher bond, header bond, stretcher/header bond and honeycomb bond brickwork, reinforced brickwork etc
Stone block work eg: masonry; rubble stone walling; stone sett and cobble paving
Accessories

Blocks, bricks
Types eg:
Hollow blocks (pots), angle blocks, wedge shape blocks
Bricks eg: common bricks including flettons; facing bricks; engineering bricks; airbricks (ventilation bricks); firebricks; solid, hollow, cellular and perforated bricks; special shape bricks; wire cut and pressed bricks
Ashlars, cobbles, setts, rubble

G LARGE BLOCK, PANEL WORK
Work with large blocks and panels

Large blocks, panels

	H	Section work	**H/M**
	I	Pipe work	
	J	Wire work, mesh work	
	K	Quiltwork	
	L	Flexible sheet work (proofing)	
	M	Malleable sheet work	

H SECTION WORK

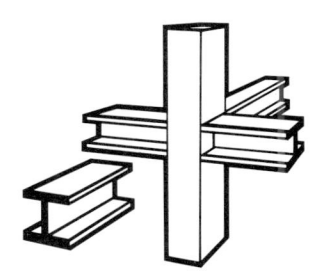

Work with sections, including hollow sections
Pipe work see I
Wire work see J
Types eg:
Structural section work; non-structural section work; sheathing, cladding section work
Accessories

Sections
Types eg:
Pins, rods, bars, rolls, cylinders, beads;
Strips eg: fillets, battens, scantlings, slatings, planks, boards, deal, laths, furrings, tapes;
Special cross sections eg: angles, channels, T sections, Z sections, rectangular hollow sections, tubes, profiles

I PIPE WORK

Work with pipes and channels, for conveying matter such as gases, liquids, cables
Types eg:
Pipe work eg: screw jointed, solder jointed pipework; duct work; trunking work
Accessories

Pipes
Types eg:
Pipes eg: hose pipes, composite pipes, junction pipes, taper pipes, pipe bends, suction pipes, flush pipes, circulation pipes, rising pipes, main pipes
Ducts, conduits
Channels eg: chutes, gutters

J WIRE WORK, MESH WORK

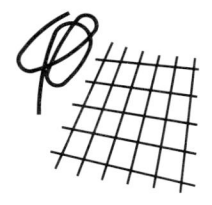

Work with wires, mesh
Types eg:
Wire work eg: cable work
Mesh work eg: mesh lathing work, wire netting backing work
Accessories

Wires, meshes
Types eg:
Wires eg: yarns, cords, ropes, strings, lines, cables, chains;
Meshes eg: netting, lattices, mesh lathing, expanded mesh, chain link

K QUILTWORK

Work with quilts
Flexible sheet work see T
Types eg:
Thatching work
Insulating quiltwork
Accessories

Quilts
Types eg:
Lagging quilts

L FLEXIBLE SHEET WORK (PROOFING)

Work with flexible sheets, strips, tiles etc for proofing
Flexible sheet work other than proofing see T
Types eg:
Sheet membrane work eg: embedded membrane work, membrane work supported at intervals;
Sheet covering work eg: single- and multi-layer work
Accessories

Flexible sheets (proofing)
Sheets for proofing which can be stored in rolls, but are used unrolled
Types eg:
Damp proof sheets, strips, tiles, foils
Vapour proof sheets, strips, foils
Foils, papers, felts including breather type

M MALLEABLE SHEET WORK

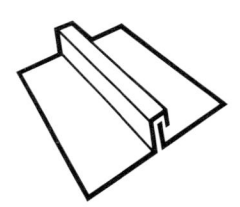

Work with malleable sheets, strips, tiles etc
Types eg:
Sheet covering work;
Sheet flashings work
Accessories

Malleable sheets
Sheets which can be bent to a tight radius without fracture and which will keep their shape when formed
Types eg:
Damp proof sheets, strip;
Protective sheets, strip

N/V

N	Rigid sheet overlap work
P	Thick coating work
R	Rigid sheet work
S	Rigid tile work
T	Flexible sheet work
V	Film coating and impregnation work

N RIGID SHEET OVERLAP WORK

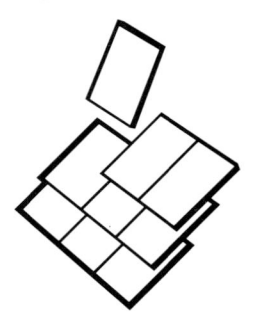

Work with rigid sheets, or tiles, overlapping each other
Types eg:
Sheet covering, cladding work eg: corrugated, profiled, troughed sheet work;
Tiling work eg: double-lap (plain), single-lap tiling work
Shingling work
Slating work
Accessories

Rigid sheets for overlapping
Types eg:
Corrugated, profiled, troughed sheets
Plain, corrugated, single-lap, double-lap tiles
Shingles
Slates

P THICK COATING WORK

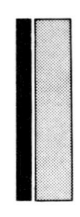

Work with thick coatings, too stiff to be applied by brush
Mesh lathing work see J
Types eg:
Screeding work; paving coating work;
Insulating coating work;
Tanking, waterproofing coating work;
Protective coating work;
Lining coating work;
Three coat work;
Two coat work;
Special surface coating work eg: roughcast, pebbledash, Tyrolean;
Rendering, backing, pricking-up, coating work;
Floating, browning, topping coating work;
Finishing, setting, skimming, fining coating work
Accessories

Q Vacant

R RIGID SHEET WORK

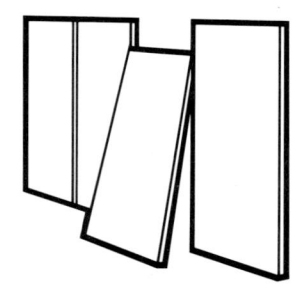

Work with rigid sheets
Rigid sheet overlap work see N
Types eg:
Sheet glazing work;
Sheet lining work;
Sheet infilling work;
Sheet cladding, casing work;
Sheet insulating work;
Sheet decking, walling work;
Accessories

Rigid sheets
Types eg:
Panes, panels, boards, plates, thin slabs;
Laminates, sandwich panels, ribbed sheets

S RIGID TILE WORK

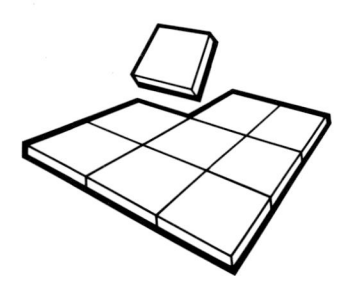

Work with rigid tiles, usually laid butt jointed
Rigid tile overlap work see N
Types eg:
Slab paving work;
Mosaic work;
Tiling work eg: parquet work
Accessories

Rigid tiles
Types eg:
Flags, slabs, tiles, tesserae

T FLEXIBLE SHEET WORK

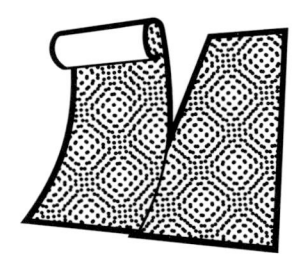

Work with flexible sheets, strips, tiles
Flexible proofing sheet work see L
Types eg:
Sheet covering work eg: flexible sheet finishings work, carpeting work, veneering work, papering work
Accessories

Flexible sheets
Sheets (and tiles cut from them) which can be stored in rolls, but are used unrolled
Types eg:
Finishing sheets, tiles;
Decorative sheets, tiles;
Papers, wallpapers, fabrics, laminates;
Veneers;
Carpets;
Ribbed sheets

U Vacant

V FILM COATING AND IMPREGNATION WORK

Work with film coatings, and impregnation work
Thick coating work see P
Types eg:
Waterproofing coating, impregnation work;
Protective coating, impregnation work;
Decorative coating, paint work, impregnation work;
Priming/coating work;
Undercoating work;
Finishing, top coating work
Accessories

W Planting work
X Work with components
Y Formless work
Z Joints

W PLANTING WORK

Work with plants and seeds
Types eg:
Cultivation work (preparatory work);
Seeding work;
Turfing work;
Planting work;
Tree work;
Rock work
Accessories

Plants, seeds
Types eg:
Trees eg: forest, woodland, timber trees; ornamental trees; fruit trees
Shrubs eg: wild scrub, bushes; ornamental, flowering shrubs; hedge plants, climbers; fruit bushes
Non-woody plants eg: perennials (herbaceous); annuals, biennials; alpines, rock plants; bulbs, corms, tubers; water and waterside plants; ferns; ornamental grasses, vegetables
Special plants eg: unusual, contorted, indoor, hothouse plants
Grasses
Seeds

X WORK WITH COMPONENTS

Work with complex components usually characterised primarily by their function

Components
Physical forms other than F to W, Y
Specific components usually coded primarily by their function in Table 1 eg:
(31)Xh complex metal components (windows/doors) for external openings
X may also be used with classes from Table 3 only eg:
Xt7 Ironmongery

Y FORMLESS WORK

Formless work eg: making mixes, not related to a specific type of construction from A to X
Types eg:
Mixes, foams, emulsions, gels, pastes, slurries;
Fluids, gases, liquids;
Amorphous solids, waxes;
Granular solids, particles, granules, chips

PRODUCTS

Z JOINTS

Joints special to classes A to X are usually included with the appropriate class but may be included here if described separately
Types eg:
Corner, heading, butt, T, edge, angle, lapped, interlocking joints
Isolation, shear resistant, expansion, contraction, movement joints
Fixed, semi-rigid, hinged, sliding joints
Soldered, welded joints
Welted, seamed joints
Mitred, mortice and tenon, combed, dowelled, saddle, tongued and grooved, dovetail, rebated, flanged, riveted, spigot and socket, screwed joints
Fish plate, gusset plate, connector joints
Concave, V, weathered, flush, raked, squeezed, struck, beaded joints

Table 3

Materials

Key

Scope
Materials which combine to form the products in Table 2*.

Structure
Two main divisions:
Building materials by substance code e to s;
Building materials by purpose coded t to w.
Note that codes a, b, c, d may be used respectively for classes (A), (B), (C) and (D) from Table 4.
When this is done Table 3 can be used as a classification of all the resources required for project construction, including management, plant and labour as well as materials*.

Changes
Most changes from the last edition are at a detailed level. It is no longer recommended that Table 3 should necessarily be used with Table 2, although the SfB Agency will normally do so on published literature.

Coding
Coding is lower case letters subdivided by numerals; 1 is omitted because of possible confusion with 'one'.
Code positions marked 'vacant' may be used for any purpose in private applications but not on published literature.

Use
Where two classes seem equally relevant – this is a matter of judgement – use the first in the absence of any private rule to the contrary. When using the table for a particular purpose, consider:

Simplicity of use
Use codes only if they serve a useful purpose.

Compound subjects
Adopt rules for dealing consistently with compound subjects which relate to two or more different materials.
In cases where a very detailed systematic breakdown is required, two codes from the table can be used together eg: h2:h5 Copper in relation to galvanised steel (see Appendix 2).

Filing
The heading 'Other' at the end of each class indicates that all other aspects of the main subject may follow on at the position. Some users may prefer to ignore these headings and file this material immediately after the main heading (see Applications 2.3).

Composite materials
Frequently, two or more materials are physically or chemically combined to create a composite material. Table 3 cannot list al composites.
A composite consisting mainly of one material with relatively small proportions o others, should be placed with the dominan material.
Composites which are not specifically liste in the table, and which consist of two or more materials of similar importance shoul be placed at the end of the relevant class o classes, at 'other types'.
Sandwich slabs are usually best classified by the 'structural skin material', not the cor or the protective surface.
Composite materials generally are at y.

*See Appendix 6

Matrix

a Materials
b Vacant*
c Vacant*
d Vacant*

Formed materials e/o

e Natural stone
f Precast with binder
g Clay (dried, fired)
h Metal
i Wood
j Vegetable and animal fibres
k Vacant
m Inorganic fibres
n Rubbers, plastics etc
o Glass

Formless materials p/s

p Aggregates, loose fills
q Lime and cement binders, mortars, concretes
r Clay, gypsum, magnesia and plastics binders, mortars
s Bituminous materials

Functional materials t/w

t Fixing and jointing materials
u Protective and process/ property modifying materials
v Paints
w Ancillary materials

x Vacant*

y Composite materials*
z Substances*

*See Appendix 6

a/h

a **Materials**
e **Natural stone**
f **Precast with binder**
g **Clay (dried, fired)**
h **Metal**

a **MATERIALS***
Materials as a whole, of two or more of specific classes e/z eg: steel plus timber

For some applications, one file may hold all information on materials. If subdivision is required it will often be enough to use only the headings below for naming and coding files:

e **Natural stone**
f **Precast with binder**
g **Clay (dried, fired)**
h **Metal**
i **Wood**
j **Vegetable and animal materials**
m **Inorganic fibres**
n **Rubbers, plastics etc**
o **Glass**
p **Aggregates, loose fills**
q **Lime and cement binders, mortars, concretes**
r **Clay, gypsum, magnesia and plastics binders, mortars**
s **Bituminous materials**
t **Fixing and jointing materials**
u **Protective and process/property modifying materials**
v **Paints**
w **Ancillary materials**
y **Composite materials**
z **Substances**

For an intermediate stage, where two or more files are needed, they may divide the schedule between them in the most convenient way eg:
e/o Formed materials
p/s Formless materials
t/w Functional materials
y/z Composite materials; substances

b **Vacant***

c **Vacant***

d **Vacant***

Formed materials e/o

e **NATURAL STONE**
Cast stone see f3
Mineral fibres see m1
Natural aggregates see p1
If necessary, subdivide eg:
e1 Granite, basalt, other igneous
e2 Marble
e3 Limestone (other than marble)
e4 Sandstone, gritstone
e5 Slate
e9 Other natural stone

f **PRECAST WITH BINDER**
Aggregates see p
Binders see q, r
Concrete, mortar mixes and in general see q, r
If necessary, subdivide eg:
Precast concrete f1/f5
f1 Sandlime concrete (precast)
Glass fibre reinforced calcium silicate (grcs)
Sandlime mixes see q1
f2 All-in aggregate concrete (precast)
Heavy concrete (precast)
Glass fibre reinforced cement (grc)
Concrete, mortar mixes and in general see q4
f3 Terrazzo (precast)
Granolithic (precast)
Terrazzo, granolithic mixes and in general see q5
Cast stone
artificial stone, reconstructed stone
Natural stone see e
Natural aggregates see p1
f4 Lightweight cellular concrete (precast)
Lightweight cellular concrete mixes and in general see q6
f5 Lightweight aggregate concrete (precast)
Lightweight aggregate concrete mixes and in general see q7
f6 Asbestos based materials (preformed) eg:
asbestos cement, asbestos-silica-cement, asbestos-silica-lime, bitumen bonded asbestos cement
Asbestos fibre see m2
Asbestos based mixes and in general see q9
f7 Gypsum (preformed)
Glass fibre reinforced gypsum (grg)
Gypsum mixes and in general see r2
f8 Magnesia materials (preformed)
Magnesia mixes and in general see r3
f9 Other materials precast with binder
Preformed asphalt see n1

g **CLAY (DRIED, FIRED)**
Clay soils see p1
Expanded clay aggregates see p3
If necessary, subdivide eg:
g1 Dried clay eg:
adobe, cob, pisé de terre
g2 Fired clay, vitrified clay, ceramics
Unglazed fired clay eg:
terra cotta, earthenware, stoneware
g3 Glazed fired clay eg:
salt-glazed ware, faience (glazed terra cotta), glazed stoneware, vitreous china, porcelain
g6 Refractory materials eg:
fireclay (alumina-silica); other refractory materials, based on silica, magnesite, carbon, silicon carbide
g9 Other dried or fired clays

h **METAL**
If necessary, subdivide eg:
Ferrous metals h1/h3
h1 Cast iron
Wrought iron, malleable iron
h2 Steel, mild steel
h3 Steel alloys eg:
stainless steel

Non-ferrous metals h4/h9
h4 Aluminium, aluminium alloys eg:
aluminium silicon, aluminium zinc, duralumin
h5 Copper
h6 Copper alloys eg:
bronze, gunmetal (copper, tin); phosphor bronze (copper, tin, phosphorus); aluminium bronze (copper, aluminium); silicon bronze (copper, silicon); beryllium bronze (copper, beryllium); nickel bronze (copper, nickel); nickel silver, german silver (copper, nickel, zinc); brass (copper, zinc)
h7 Zinc
h8 Lead, white metal
h9 Chromium, nickel, gold; other metals, metal alloys

See Appendix 6

		i	**Wood**		**i/o**
		j	**Vegetable and animal materials**		
		m	**Inorganic fibres**		
		n	**Rubbers, plastics etc**		
		o	**Glass**		

i WOOD
including wood laminates
Wood fibre materials see j1
Wood particle materials see j7
Wood wool-cement see j8
If necessary, subdivide eg:

- i1 Timber (unwrot)
- i2 Softwood (in general, and wrot) eg:
  fir, hemlock, spruce, whitewood, larch, cedar, pine, redwood
- i3 Hardwood (in general, and wrot) eg:
  walnut, beech, oak, elm, iroko, teak, mahogany, sapele
- i4 Wood laminates eg:
  plywood, laminboard, blockboard; with faces or veneers of wood, metal, plastics
- i5 Wood veneers
- i9 Other wood materials, except those at j1, j7, j8

j VEGETABLE AND ANIMAL MATERIALS
including fibres and particles and materials made from these
Wood see i
Inorganic fibres see m
Natural rubber see n5
Synthetic fibres see n6
Mixed fibres see n9
If necessary, subdivide eg:

- j1 Wood fibres eg:
  fibre building board, hardboard, medium board
- j2 Paper eg:
  kraft paper, corrugated paper, metal backed paper, bituminised paper; cardboard, pulpboard, pasteboard, millboard
- j3 Vegetable fibres other than wood eg:
  grasses (bamboo, eelgrass, sea grass, reeds, rushes, straw, strawboard);
  stem fibres (hemp, jute, flax, linen, flaxboard, hessian);
  leaf fibres (sisal); seed hairs (cotton, coir)
- j5 Bark, cork
- j6 Animal fibres eg:
  hair, wool, leather
- j7 Wood particles eg:
  chipboard
- j8 Wood wool-cement
- j9 Other vegetable and animal materials
  Linoleum (hessian-linseed oil) see n4

k Vacant

l Vacant

m INORGANIC FIBRES
including materials made from these
Vegetable and animal fibres see j
Synthetic fibres see n6
Mixed fibres see n9
If necessary, subdivide eg:

- m1 Mineral wool/fibres eg:
  vermiculite
  Natural stone see e
  Aggregates, loose fill see p
  Glass wool/fibres
  Glass see o
  Ceramic wool/fibres
  Ceramics see g2
- m2 Asbestos wool/fibres
  Asbestos based materials see f6 (preformed) and q9
- m9 Other inorganic fibrous materials eg:
  carbon fibres

n RUBBERS, PLASTICS ETC
If necessary, subdivide eg:

- n1 Asphalt (preformed)
  Asphalt in general see s
- n2 Impregnated fibre and felt eg:
  bituminous felt, bitumen polythene, pitch fibre
  Bitumen bonded asbestos cement see f6
  Bituminised paper see j2
- n4 Linoleum

Synthetic resins n5/n6
- n5 Rubbers (elastomers) eg:
  natural rubber (polyisoprene);
  synthetic rubbers including polychloroprene (neoprene);
  butyl, polyisobutylene;
  polysulphides;
  chlorosulphonated polyethylene, ethylene propylene;
  nitrile; polybutadiene;
  ebonite (polyisoprene); silicone;
  urethane; chlorinated rubber
- n6 Plastics, including synthetic fibres
  Natural fibres see j, m
  Two part synthetic resin mortars see r4
  Thermoplastics eg:
  polyvinyl chloride (PVC);
  polyvinyl acetate (PVA);
  polyethylene (polythene);
  polypropylene; polystyrene;
  acrylic resins;
  cellulose derivatives (acetate rayons, viscose rayon);
  coumarone resins;
  acetal resins;
  polyamides (nylon);
  polyesters
  Thermosets eg:
  phenolic resins;
  melamine resins;
  urea resins;
  polyesters;
  epoxy resins;
  polyurethanes
- n7 Cellular plastics
  foamed plastics;
  expanded plastics
  including types listed at n6
- n8 Reinforced plastics eg: grp plastics laminates
  including types listed at n6
- n9 Other rubber, plastics materials eg:
  mixed natural/synthetic fibres

o GLASS
Glass fibre see m1
If necessary, subdivide eg:

- o1 Clear, transparent, plain glass eg:
  coloured
- o2 Translucent glass eg:
  coloured, patterned, textured, diffuse reflecting
- o3 Opaque, opal glass eg:
  coloured
- o4 Wired glass
- o5 Multiple glazing
- o6 Heat absorbing/rejecting glass, X-ray absorbing/rejecting glass, solar control glass
- o7 Mirrored glass, 'one way' glass, anti-glare glass
- o8 Safety glass, toughened glass, laminated glass, security glass, alarm glass
- o9 Other glass, including cellular glass

p/s

p Aggregates, loose fills
q Lime and cement binders, mortars, concretes
r Clay, gypsum, magnesia and plastics binders, mortars
s Bituminous materials

Formless materials p/s

p AGGREGATES, LOOSE FILLS
Natural stone see e
Mineral fibres see m1
If necessary, subdivide eg:

p1 Natural fills, aggregates eg:
soils, earth, including clay soils, sandy soils, pebbly soils, diatomite (molar earth, kieselguhr); crushed stone, gravel, sand, hoggin, ballast, shingle

p2 Artificial aggregates in general eg:
artificial granular aggregates (heavy), including air-cooled blast furnace slag, crushed concrete, crushed brick, hardcore

p3 Artificial granular aggregates (light) eg:
foamed blast furnace slag; clinker; breeze; exfoliated vermiculite; expanded mica and perlite; expanded clays, shales and slates; sintered pulverised fuel ash

p4 Ash eg:
fly ash, pulverised fuel ash

p5 Shavings
p6 Powder
p7 Fibres
p9 Other aggregates, loose fills

q LIME AND CEMENT BINDERS, MORTARS, CONCRETES
Precast with binder see f
Admixtures see u2
If necessary, subdivide eg:

q1 Lime (calcined limestones)
hydrated lime, lime putty, lime-sand mix (coarse stuff)
Mineral limestones see e2, e3

q2 Cement, hydraulic cement eg:
portland cement, including rapid hardening, low heat, sulphate resisting, white, coloured, blast furnace, slag cement; supersulphated cement; high alumina cement, ciment fondu; masonry cement, oil well cement; cement-sand mix

q3 Lime-cement
mixed hydraulic, semi-hydraulic, non-hydraulic binders

q4 Lime-cement-aggregate mixes
Mortars (ie with fine aggregates) eg:
lime mortars, lime plasters; cement mortars, cement plasters; cement-lime mortars, compo mortars
Gypsum plaster see r2
Synthetic resin mortars see r4
Concretes (ie with fine and/or coarse aggregates) eg:
'all-in' aggregate concrete; heavy concrete; plain concrete, mass concrete; reinforced concrete; prestressed concrete
Precast concrete see f1/f5

q5 Terrazzo (mixes and in general); Granolithic (mixes and in general) including special aggregates such as aluminium oxide, carborundum, metal, flint, quartz

q6 Lightweight, cellular concrete (mixes and in general)

q7 Lightweight aggregate concrete (mixes and in general)

q9 Other lime-cement-aggregate mixes eg:
asbestos cement mixes (sprayed asbestos)
Preformed asbestos based materials see f6
Asbestos fibre see m2
Wood wool-cement see j8

r CLAY, GYPSUM, MAGNESIA AND PLASTICS BINDERS, MORTARS
If necessary, subdivide eg:

r1 Clay mortar mixes; refractory mortar
Refractory materials in general see g6

r2 Gypsum
alabaster, hemihydrate, anhydrite;
Gypsum mixes, gypsum plasters
plaster of paris, retarded hemihydrate, anhydrous gypsum plaster, Keene's cement, Parian cement, Martin's cement, Mack's cement

r3 Magnesia
Magnesia mixes, magnesium oxychloride, magnesium oxysulphate

r4 Plastics binders
Plastics mortar mixes
synthetic resin compounds and mortar mixes; mixes of aggregates and plastics binders; cement-synthetic resin emulsions, including cement-rubber latex, cement-neoprene latex, cement-PVC or PVA emulsions
Plastics in general see n6
Adhesives in general see t3

r9 Other binders and mortar mixes

s BITUMINOUS MATERIALS
Asphalt (preformed) see n1
Impregnated fibre and felt see n2
Adhesives in general see t3
Protective materials in general see u
If necessary, subdivide eg:

s1 Bitumen
natural bitumen, petroleum bitumen, bitumen solution, bitumen emulsion, bitumen paints;
tar, coal tar;
pitch, coal tar pitch;
asphalt, lake asphalt

s4 Mastic asphalt (fine or no aggregate)
pitch mastic

s5 Clay-bitumen mixes; stone-bitumen mixes (coarse aggregate)
Rolled asphalt; coated macadam
bitumen macadam, tar macadam

s9 Other bituminous materials

t **Fixing and jointing materials**
u **Protective and process/property modifying materials**
v **Paints**
w **Ancillary materials**
y **Composite materials**
z **Substances**

t/z

Functional materials t/w

t **FIXING AND JOINTING MATERIALS**
If necessary, subdivide eg:
- t1 Welding materials
- t2 Soldering materials
- t3 Adhesives, bonding materials eg:
  natural adhesives including animal and vegetable glues;
  synthetic adhesives;
  gap-filling adhesives;
  close contact adhesives
- t4 Joint fillers eg:
  putty; mastics; sealants; gaskets

Ironmongery t6/t7
- t6 Fasteners, 'builders ironmongery'
  Anchoring devices eg:
  fixing slips, plugs
  Attachment devices eg:
  clips, clamps, cleats, connectors, couplings, hinges
  Fixing devices eg:
  rivets, nails, nuts, bolts, screws, studs; couplings
- t7 'Architectural ironmongery'
- t9 Other fixing and jointing agents

u **PROTECTIVE AND PROCESS/PROPERTY MODIFYING MATERIALS**
If necessary, subdivide eg:
- u1 Anti-corrosive materials, treatments
  Paints see v
  Metallic coatings applied by eg:
  electroplating, including anodising, galvanising, nickel plating, chromium plating;
  electrostatically charging;
  cementing, including sherardising, hot-dipping; spraying
  Non-metallic coatings applied by eg:
  chemical conversion, including oxide film formation, phosphating
- u2 Modifying agents, admixtures eg:
  setting agents, curing agents including setting accelerators, setting retardants;
  hardeners;
  drying agents; wetting agents;
  water retaining agents;
  frost resisting agents;
  dispersing agents;
  solvents; water softeners;
  foaming agents including air-entraining agents;
  anti-foaming agents;
  emulsifying agents;
  catalysts;
  accelerators; retardants;
  stabilisers

  Workability aids
  including anti-skinning agents;
  anti-settling agents, suspending agents;
  plasticisers; extenders;
  thinners; flocculants, coagulants, gelling agents;
  water reducing agents;
  flatting agents;
  cross-linking agents
- u3 Materials resisting special forms of attack such as rot, fungus, insects, condensation eg:
  rotproofers, fungicides, insecticides, preservatives including tar oil preservatives (creosote), organic solvent preservatives, water-borne preservatives
- u4 Flame retardants if described separately
- u5 Polishes; seals; surface hardeners including treatments such as nitriding; size
- u6 Water repellents if described separately
- u9 Other protective and process/property modifying agents eg:
  ultra-violet absorbers;
  anti-static agents

v **PAINTS**
Protective materials in general see u
If necessary, subdivide eg:
- v1 Stopping, fillers, knotting, paint preparation materials including primers
- v2 Pigments, dyes, stains
- v3 Binders, media eg:
  drying oils including linseed oil, tung oil, castor oil
- v4 Varnishes, lacquers eg:
  resins including copal, shellac
  Enamels, glazes
- v5 Oil paints, oil-resin paints, synthetic resin paints;
  complete systems including primers
- v6 Emulsion paints where described separately
  synthetic resin-based emulsions;
  complete systems including primers
- v8 Water paints eg:
  cement paints;
  whitewash, distempers
- v9 Other paints eg:
  metallic paints, paints with aggregates

w **ANCILLARY MATERIALS**
If necessary, subdivide eg:
- w1 Rust removing agents
- w3 Fuels eg:
  gas, liquid fuels, solid fuels
- w4 Water
- w5 Acids, alkalis
- w6 Fertilisers eg:
  inorganic, organic including bulky organic manures;
  vegetative products eg:
  straw, moss, fibre;
  soil conditioners, growth stimulants
- w7 Cleaning materials eg:
  detergents, soaps
  Abrasives
- w8 Explosives
- w9 Other ancillary materials eg:
  weedkillers, insecticides, fungicides, growth inhibitors

x **Vacant**

y **COMPOSITE MATERIALS**
See note in key to Table 3

z **SUBSTANCES**
If necessary, subdivide eg:
- z1 By state eg:
  solids; fluids; mixtures of states
- z2 By chemical composition eg:
  inorganic materials, organic materials
- z3 By origin eg:
  naturally occurring materials;
  manufactured materials
- z9 Other substances

Table 4

Activities, requirements

Key

Scope
Tables 0, 1, 2 and 3 identify objects which are physically incorporated in, or result from, the building process. Table 4 identifies other factors relating to building ie objects which assist or affect construction but are not incorporated in it, and factors such as activities, requirements, properties, processes, etc.

Structure
Two main divisions:
Construction activities, plant, tools, etc, coded (A) to (D);
Properties, processes, etc, coded (E) to (Y).
The table also includes codes at (Z) which make it possible to relate CI/SfB to subjects which lie outside the construction industry eg law, medicine, philosophy, also space, time.
A diagrammatic view of the whole table is shown on the next page.

Changes
Main classes from (E) onwards in CI/SfB 1968 were based on the 1964 edition of the CIB Master List. A new edition of the Master List superseded the 1964 list in 1972, and this version is followed in Table 4 of this edition of CI/SfB.
A substantial number of changes have had to be made to main classes – as well as to class subdivisions – from (E) onwards to keep the two documents in step.
Appendix 6 shows the relation between main headings and codes in CI/SfB and the CIB Master List.

Coding
Coding is capital letters in brackets. (I) and (O) are omitted because of possible confusion with 'one' and 'nought'.
Code positions marked 'vacant' may be used for any purpose in private applications but not on published literature.

Use
Where two classes seem equally relevant – this is a matter for personal judgement – use the first in the absence of any private rule to the contrary. When using the table for a particular use, consider:

Simplicity of use
For a small set of data it may be possible to use the simple alphabetical codes and ignore all the detailed codes and headings. Use codes only if they serve a useful purpose.

Compound subjects
Adopt rules for dealing consistently with compound subjects which relate to two or more different classes.
In cases where a very detailed systematic breakdown is required, two codes from the table can be used together eg (F4) (M) Heat in relation to dimensions (eg effect of heat on dimensions).
'Other' at the end of each class indicates that all other aspects of the main subject may follow on at the position.
Some users may prefer to ignore these headings and file this material immediately after the main heading (see Applications 2.3).

Table 4 matrix

Activities, aids (A)/(D)

(A)	Administration, management activities, aids*	(A1)	Organizing offices, projects	(A2)	Financing, accounting	(A3)	Designing, physical planning	(A4)	Cost planning, cost control; tenders, contracts
(B)	Construction plant, tools*	(B1)	Protection plant	(B2)	Temporary (non-protective) works	(B3)	Transport plant	(B4)	Manufacture, screening, storage plant
(C)	Vacant*								
(D)	Construction operations*	(D1)	Protecting	(D2)	Clearing, preparing	(D3)	Transporting, lifting	(D4)	Forming: cutting, shaping fitting

Requirements, Properties processes (E)/(Y)

Description (E)/(G)

(E)	Composition etc	(E1)	Content	(E2)	Production	(E3)	Manufacture	(E4)	On-site assembly
(F)	Shape, size etc	(F1)	Shape	(F2)	Vacant	(F3)	Vacant	(F4)	Size
(G)	Appearance etc	(G1)	Appearance	(G2)	Vacant	(G3)	Vacant	(G4)	Texture, pattern
(H)	Context, environment	(H1)	Climate, ecological	(H2)	Modified environments	(H3)	Vacant	(H4)	Amenities

Performance (J)/(T)

(I)	Vacant								
(J)	Mechanics	(J1)	Mechanics of structure, soil	(J2)	Mechanics of fluids	(J3)	Strength	(J4)	Loads, forces
(K)	Fire, explosion	(K1)	Sources, types	(K2)	Fire protection	(K3)	Fire resistance (structures)	(K4)	Reaction to fire (materials)
(L)	Matter	(L1)	Fluids	(L2)	Gases, vapours	(L3)	Liquids	(L4)	Solids
(M)	Heat, cold	(M1)	Sources, types	(M2)	Insulation	(M3)	Transfer	(M4)	Capacity
(N)	Light, dark	(N1)	Sources, types	(N2)	Insulation	(N3)	Transfer	(N4)	Reflection, refraction
(P)	Sound, quiet	(P1)	Sources, types	(P2)	Insulation	(P3)	Transfer	(P4)	Reflection
(Q)	Electricity, magnetism, radiation	(Q1)	Sources, types	(Q2)	Insulation	(Q3)	Transfer	(Q4)	Ionisation
(R)	Energy, other physical factors	(R1)	Energy sources, types	(R2)	Potential energy	(R3)	Demand, transfer of energy	(R4)	Control of energy
(S)	Vacant								
(T)	Application	(T1)	Types of use	(T2)	Proper use	(T3)	Efficiency	(T4)	Utilization, obsolescence
(U)	Users, resources	(U1)	Communities	(U2)	Society, custom	(U3)	People	(U4)	Physical, mental factors
(V)	Working factors	(V1)	Ease of storing	(V2)	Demountability	(V3)	Ease of moving	(V4)	Ease of cutting
(W)	Operation, maintenance, factors	(W1)	Operation	(W2)	Routine servicing	(W3)	Vacant	(W4)	Vacant
(X)	Change, movement, stability factors	(X1)	Integration etc	(X2)	Disintegration etc	(X3)	Transfer, movement	(X4)	Considerable change
(Y)	Economic, commercial factors	(Y1)	Availability of finance	(Y2)	Costs, prices	(Y3)	Initial capital	(Y4)	Cost in use, running costs
(Z)	Peripheral subjects, form of presentation, time, place								

* see Appendix 6

Table 4 matrix

(A5) Production planning, progress control	(A6) Buying, delivery	(A7) Inspection, quality control	(A8) Handing over, feedback, appraisal	(A9) Other activities: arbitration, insurance
(B5) Treatment plant	(B6) Placing, pavement, compaction plant	(B7) Hand tools	(B8) Ancillary plant	(B9) Other construction plant, tools
(D5) Treatment: drilling, boring	(D6) Placing: laying, applying	(D7) Making good, repairing	(D8) Cleaning up	(D9) Other construction operations

(E5) Automation	(E6) Vacant	(E7) Connection, fixing	(E8) Accessories	(E9) Other
(F5) Weight	(F6) Tolerance, accuracy	(F7) Dimensional systems	(F8) Other	
(G5) Colour	(G6) Opacity	(G7) Other visual	(G8) Feel, smell, aural, taste, other	
(H5) Vacant	(H6) Internal environments	(H7) Occupancy	(H8) Other	

(J5) Adhesion	(J6) Vibration	(J7) Deformation, stability	(J8) Other
(K5) Smoke etc	(K6) Explosions	(K7) Fire, explosion damage, salvage	(K8) Other
(L5) Physical, chemical	(L6) Biological	(L7) Pollution	(L8) Deterioration
(M5) Heat gain, loss	(M6) Overheating, inadequate heat	(M7) Temperature	(M8) Control, other
(N5) Brightness	(N6) Glare	(N7) Illumination, candlepower	(N8) Control, other
(P5) Sound gain, loss	(P6) Loudness	(P7) Measurement of sound	(P8) Control, other
(Q5) Voltage, capacitance	(Q6) Magnetism, electro-magnetism	(Q7) Radiation	(Q8) Control, other
(R5) Other energy factors	(R6) Side effects	(R7) Compatability	(R8) Durability

(T5) Change in use, re-use	(T6) Conservation of resources, waste	(T7) Mis-use	(T8) Failure in use, defects
(U5) Animals, plants, machines	(U6) Resources	(U7) Vacant	(U8) Other
(V5) Ease of drilling	(V6) Ease of placing, fixing	(V7) Adaptability	(V8) Ease of disposing
(W5) Repair	(W6) Alteration	(W7) Restoration, reconstruction	(W8) Other

(X5) Stability	(X6) Gradual change	(X7) Quality change	(X8) Other

(Y5) Vacant	(Y6) Values and benefits	(Y7) Economic efficiency	(Y8) Supply factors	(Y9) Services factors
(Z5) Vacant				

(A)   (A)   Administration and management activities, aids

ACTIVITIES, REQUIREMENTS

For some applications, one file may hold all information on activities and requirements. If subdivision is required it will often be enough to use only the headings below for naming and coding files:

- **(A)** Administration and management activities, aids
- **(B)** Construction plant, tools
- **(D)** Construction operations
- **(E)** Composition
- **(F)** Shape, size
- **(G)** Appearance
- **(H)** Context, environment
- **(J)** Mechanics
- **(K)** Fire, explosion
- **(L)** Matter
- **(M)** Heat, cold
- **(N)** Light, dark
- **(P)** Sound, quiet
- **(Q)** Electricity, magnetism, radiation
- **(R)** Energy, side effects, compatibility, durability
- **(T)** Application
- **(U)** Users, resources
- **(V)** Working factors
- **(W)** Operation, maintenance factors
- **(X)** Change, movement, stability factors
- **(Y)** Economic, commercial factors
- **(Z)** Peripheral subjects; form of presentation; time; space

For an intermediate stage, where two or more files are needed, they may divide the schedule between them in the most convenient way eg:

- (A)   Administration
- (B)/(D)   Plant and operations
- (E)/(Y)   Requirements, properties
- (Z)   Peripheral subjects

Activities, aids (A)/(D)

(A)   **ADMINISTRATION AND MANAGEMENT ACTIVITIES, AIDS***

Classes f to q can be used to subdivide any or all of the activity classes from (A1) to (A9).
They should be used with (A) for the classification of documents dealing with construction industry subjects in a general way, not related to project, or office, administration and management eg:
- (Af)   administration, organization, but
- (A1f)   administration, organization of offices, projects

(A1) should be used for the classification of all information on administration relating to the founding and establishing of projects (and the offices controlling projects), unless a more precise classification can be given under (A2) to (A9). Equipment and documentation required for any activity should be classified with the activity whenever possible.
Any class below can be sub-divided by any previous class by the direct addition of the lower case letter eg:
- (Ajfh)   growth, change in controls

(Af)   **Administration, organization**
Administrative structure in general
including co-operation, co-ordination, communication, cybernetics, systems theory, organization of ideas, etc.
Administration relating directly to project activities (and offices controlling projects) see (A1) onwards

- **(Afg)**   Objectives, policy in general
- **(Afh)**   Growth, changes in general
- **(Afi)**   Decision making in general
- **(Afk)**   Human relations in general
  labour, industrial relations
  Personnel in general see (Am)
- **(Afn)**   Conduct, responsibility, ethics in general
  Codes of conduct classify primarily by organizations, see (Ajw), (Ak)
- **(Afp)**   Competition in general
  competitions

(Ag)   **Communications**
including research into recording, organizing, presenting and disseminating information, knowledge

- **(Agh)**   Terminology, nomenclature, identification, glossaries, etc
- **(Agi)**   Media of communication, Audio-visual communication in general
- **(Agj)**   Face to face, direct communications; meetings, committees in general
- **(Agk)**   Recorded communications in general
- **(Agm)**   Documentation, information storage and retrieval, bibliography, sources of information
  Project documentation see (A3)
- **(Agn)**   Mechanization in information, data processing in general
- **(Ago)**   Computers, calculators in general

(Ah)   **Preparation of documentation, trade and technical literature etc**
including information documentation

- **(Ahi)**   Classification, coding, indexing, terminology in general
- **(Ahk)**   Photography if not in (A1t)
- **(Ahm)**   Typography, graphic design
- **(Ahp)**   Publishing
- **(Ahq)**   Printing, lithography
  Office copying see (A1t)

(Ai)   **Public relations, publicity**
including advertising

---

* See Appendix 6

(A) Administration and management activities, aids

(Aj) Controls, procedures
Classify primarily by subject whenever possible
Organizations in general see (Ak)
Controls relating directly to project activities (and offices) see (A1) onwards

- **(Ajk)** Legislation, building case law in general
including Acts of Parliament, contract, tort
- **(Ajm)** Memoranda, orders based on legislation
- **(Ajn)** Regulations, bye-laws
see also (A3j)
- **(Ajo)** Licences
- **(Ajp)** Official advice, standards in general,
including those which are referred to in legislation, memoranda etc
Standardization see (Ap)
- **(Ajr)** British Standard Specifications Codes of Practice
- **(Ajs)** Agrément certificates
- **(Ajt)** BRE digests
- **(Aju)** DoE advisory leaflets
- **(Ajv)** Other advisory publications, standards etc in general
including those issued by trade associations
- **(Ajw)** Charters, codes of conduct, etc in general
See also (Afn) and (Ak)
- **(Ajx)** Patents, copyright in general
- **(Ajy)** Public enquiries, legal proceedings etc; appeals in general
Arbitration see (A9)

(Ak) Organizations
Classify organizations primarily by function whenever possible, eg:
research organizations under research
- **(Akm)** Government and related organisations
- **(Akn)** Central government
ministries, departments, related organizations
Local government
- **(Ako)** Commercial organizations
- **(Akp)** Other permanent organizations
including professional bodies, trade unions, associations classify primarily by subject where possible, eg: ICE under civil engineering works
- **(Akq)** Organizations for events, conferences, exhibitions etc
- **(Akr)** Design and construction organizations in general
- **(Aks)** Public: central government architect's departments etc. local government engineer, surveyor, architect, planning departments including direct works organizations, consortia
- **(Akv)** Private: partnerships, group practices, etc
- **(Aky)** Commercial: construction companies, package dealers, 'all-in' service etc architects in industry

(Am) Personnel, skills, professions, roles
Subdivide the classes below using A/Z
classify personnel according to subject whenever possible, eg: computer specialists under computers
- **(Amt)** Client
- **(Amu)** Manufacturer
- **(Amv)** Design and construction (or neither) eg:
project managers, technicians
- **(Amw)** Mainly design (design team, designer):
architects, artists, industrial designers, landscape architects, planners, quantity surveyors, services engineers, structural engineers etc
- **(Amx)** Mainly construction (construction team, constructor, builder)
Code (C) may be used for a breakdown of construction personnel

(An) Education, training, retraining (for the construction industry)
including research if not described separately
Training relating directly to projects and offices see (A1) onwards
- **(Ano)** Schools of architecture, schools of building etc
- **(Anp)** Courses, conferences
- **(Anq)** Study tours

(Ao) Research, development
Research, development relating directly to project activities (and offices) see (A1) onwards

(Ap) Standardization, rationalization
Standardization, rationalization relating directly to projects and offices see (A1) onwards
Industrialization etc see (E5)

(Aq) Testing, evaluating
Testing relating directly to projects and offices see (A1) onwards,
Quality control see (A7q)

(A)

(A)	Administration and management activities, aids
(A1)	Organizing offices, projects
(A2)	Financing, accounting

(A1) Organizing offices, projects
Overall control and organization of project activities (and therefore of offices responsible for project activities). This class should be used for the classification of all information on founding and establishing projects (and offices controlling projects) unless a more precise classification can be given under (A2) to (A9)
Equipment and documentation required for any activity should be classified with the activity whenever possible

(A1f) **Organization, organizational structure**
including co-operation, negotiation; office, project management in general; co-ordination, communication, cybernetics, systems theory, etc
Subdivide as (Af)
Organization of special activities see (A2)/(A8)

(A1fg) Objectives, policy, constitution
(A1fi) Decision taking
(A1fk) Human relations
Personnel see (A1m)
(A1fn) Conduct, responsibility, ethics
(A1fp) Means of obtaining commissions including competitions
Publicity see (A1i)
(A1fx) Co-ordination, organization, plans of work eg RIBA Plan of Work
(A1g) **Communications**
Installations see (A1s) if not in Table 1 (64)
(A1gj) Meetings; committees
(A1gm) Documentation, information storage and retrieval, information documentation, information services including library practice, equipment etc
(A1h) Preparation of documentation
(A1i) Public relations, publicity, advertising
(A1j) **Controls, procedures, legislation**
Controls, procedures for special activities see (A2)/(A8)
Contracts in general see (A4)
Contracts of service see (A1mq)
(A1jw) Professional handbooks: job manuals, office manuals, etc

(A1m) **Personnel, agreements with personnel**
Personnel for special activities see (A2)/(A8)
(A1mm) Grading
(A1mn) Selection
Advertising for staff etc
(A1mo) Applications for employment
(A1mp) Interviews
(A1mq) Engagements, appointments, agreements, partnerships
conditions of service
contracts of service
Contracts in general see (A4)
(A1mr) Personal details, welfare:
cars if not in (A2)
pensions, life assurances if not in (A2)
profit sharing, bonuses if not in (A2)
records if not in (A2)
(A1ms) Roles
Subdivide as (Am)
(A1mt) Client
(A1mu) Manufacturer
(A1mv) Designer and constructor
(A1mw) Mainly designer
(A1mx) Mainly constructor
(A1my) Authority:
directors, partners, consultants, technical staff including students, administrative staff including secretarial staff
(A1n) Staff training
(A1o) Operational Research
(A1p) O and M
Standardization, rationalization of special activities see (A2)/(A8)
Work study in general see (A5p)
(A1r) **Accommodation**
including premises, space needs in general,
fixtures and fittings if not in Table 1 (7–), (8–)
Subdivide by (E)/(Y) as necessary, eg:
(A1r)(N) lighting
(A1r)(W) maintenance
Offices of building practitioners including drawing offices (examples) see Table 0, 32
Rent, rates etc see (A2)
Insurance see (A9)
(A1s) Secretarial activities in general, including equipment, stationery
Information handling see (A1gm)
GPO telephone, telex, including procedures, dialling codes etc if not in Table 1, (64)
Internal telephones if not in Table 1, (64)
GPO post, postal requirements, etc
(A1t) Copying, photocopying
photography, microfilming etc including equipment, stationery
(A1u) Filing including equipment

(A2) Financing, accounting
Economic resources, capital etc, finance for project (and office), book-keeping
Means of meeting the cost plan, budget. May also include integrated financial/costs systems, (A2) + (A4)
Cost planning, control see (A4)
Economics in general see (Y)

(A2f) Establishing accounting system
(A2j) Controls, procedures, legislation affecting financing, accounting,
taxation, auditing
(A2r) Accounts, trading, profits etc
forecasts, budget, valuation of work in progress (if not in A4)
(A2s) Income, earnings, capital
dividends, fees, sales
(A2t) Expenditure
Indirect costs:
staff wages, salaries, pensions, etc including cost of staff and training; other outgoings on rates, stores, printing etc, purchases, rentals, cost of accommodation, depreciation, bad debts
Direct costs:
staff, labour, materials, plant
(A2u) Borrowing, lending

(A) Administration and management activities, aids
(A3)/(A9) Other aspects of organizing offices, projects

(A) (A)

(A3) **Designing, physical planning**
Investigating and physical planning including inception, feasibility, outline proposals, scheme design, detail design, production information
May also include design cost planning and production planning, (A3) + (A4) + (A5)
Physical quality control see (A7)

(A3f) Establishing methodology of physical planning, process of design, designing
Production control, allocation of resources see (A5)

(A3gj) Meetings including design team meetings

(A3h) Preparation of project documentation in general
classify types of documents with the activity which produces them whenever possible

(A3j) Controls, consents and approvals
appeals, procedures, legislation, building regulations, bye-laws

(A3r) Brief, instructions from client
ascertaining client's requirements, user requirement studies, activity data method

(A3s) Site investigation, surveying
including equipment, maps

(A3t) Drawing, drawings
including equipment

(A3u) Specifying, specifications
scheduling, schedules

(A4) **Cost planning, cost control, tenders, contracts**
Accounting see (A2)
Integrated financial/costs system see (A2)
Economics in general see (Y)

(A4f) Establishing methodology of cost planning, cost control
Costs, prices see (Y)

(A4gm) Documentation for cost planning etc

(A4gn) Mechanization, data processing

(A4go) Computers

(A4h) Preparation of documentation in general, contract documents
classify types of documents with the activity which produces them whenever possible

(A4j) Controls, procedures affecting cost planning etc
cost limits, cost yardsticks

(A4s) Quantifying
methods of measurement, Bills of Quantities

(A4t) Estimating, pricing, giving quotations, costing, price evaluation, tendering in general

(A4u) Contracts in general, nominations
Terms of agreement and conditions of contract if described separately see (A1)

(A4v) Cost analysis

(A4w) Variations, dayworks, financial control

(A4x) Certificates, valuations, fluctuations
final account if not in (A2)

(A5) **Production planning, progress control**
Site management in general
Organization of project activities (and of organizations controlling projects) in general see (A1)

(A5f) Establishing resource requirements, planning, controlling progress in general; means of fulfilling plan of work

(A5gj) Meetings including site meetings

(A5gm) Documentation for production planning etc

(A5gn) Mechanization, data processing
classify by main activity whenever possible eg: data processing for preparation of documentation (A5h)

(A5go) Computers

(A5h) Preparation of documentation
classify types of documents with the activity which produces them whenever possible

(A5j) Site manufacturing controls, procedures

(A5p) Work study in general

(A5r) Incentives, effect of delay

(A5s) Programming techniques

(A5t) Network analysis, including CPM, PERT, updating procedure

(A5u) Line of balance

(A5v) Bar (GANTT) Charts

(A6) **Buying, delivery**
Accounting see (A2)

(A6f) Organization, control of buying selling, marketing, sales forecasts, negotiations, etc

(A6gj) Meetings, visits by sales representatives etc

(A6j) Import controls

(A6o) Market research

(A6r) Receipt of orders, ordering procedure, advance ordering

(A6s) Stock control

(A6t) Distribution

(A6u) Delivery, taking delivery

(A7) **Inspection, quality control**
Cost control see (A4)
Progress control, production control in general see (A5)

(A7f) Establishing, organization of inspection, quality control

(A7j) Controls, procedures, legislation affecting inspection, quality control
Specifying see (A3u)

(A7q) Testing for quality control
commissioning projects in general
Testing, evaluating in general see Aq

(A7r) Suspect, defective work in general
Completion, defects procedure see (A8r)

(A8) **Handing over, feedback, appraisal**

(A8f) Organization of handover etc

(A8h) Building owner's manual

(A8r) Completion
maintenance inspection period, defects procedure, making good, practical completion

(A8s) Feedback

(A9) **Other activities; arbitration, insurance**

(A9r) Arbitration

(A9s) Insurances in general

(A9t) Articles, materials found on site, not connected with the project

(A9u) Contingencies

(B) Construction plant, tools*

(B) CONSTRUCTION PLANT, TOOLS*
Equipment used for construction purposes, including temporary works required for construction
Construction work (plant and operations together) see (D)
If necessary, subdivide eg:

(B1) Protection plant
- (B1a) Protection of site
  fences, guard rails
- (B1b) Temporary accommodation, site huts, etc
  canteens, latrines, site offices
- (B1c) Protection of labour
  clothing, helmets, warning notices etc
- (B1d) Protection of materials, works in progress
  air houses, alarms, containers, electric blankets
  See also (B8)
- (B1m) Other plant for protecting site

(B2) Temporary (non-protective) works
Plant for clearing and preparing site
Hand tools if described separately see (B7)
- (B2b) Demolition plant, rock blasting equipment
- (B2c) Site roads, temporary services including lighting
- (B2d) Scaffolding, shores, underpinning, formwork, casting, battery casting equipment including complete systems
  centering, forms, moulds, panels, props, runners, soldiers, struts, release agents
- (B2e) Digging, loading and excavation plant
- (B2f) Earth-moving plant in general including tractors, agricultural wheeled tractors, large crawler tractors, large rubber-tyred tractors etc
- (B2g) Tractor attachments
  angle dozers, back hoes, bull dozers, loading shovels, push dozers, rippers, shovels, tilt dozers, winches
- (B2h) Scrapers
- (B2i) Graders
- (B2j) Excavating plant in general
  face shovel, dray shovel, skimmer, drag line, drag and clam shell; also hydraulic excavators, bucket wheel excavators
- (B2k) Continuous loaders
  excavating belt conveyor loaders, bucket elevator loaders
- (B2m) Intermittent loaders
  crawler or pneumatic-tyred tractor shovels used as loaders or excavators
- (B2n) Trenchers
  ladder type boom, vertical boom, wheel type boom
- (B2p) Pile driving and extracting plant, pumps, dredgers
- (B2q) Soil testing equipment
- (B2r) Pile driving equipment
  temporary sheet piling, cofferdams, hammers, extractors, winches, pile frames, equipment for bored and in situ piling
- (B2t) Pumps, hydrophones, de-watering equipment eg:
  centrifugal, lift sludge, submersible pumps
- (B2w) Boats and dredgers
- (B2y) Other plant for clearing and preparing site

(B3) Transport plant
Hand tools if described separately see (B7)
- (B3b) Transport and haulage (mainly horizontal)
- (B3d) Trucks, lorries, prime movers, trailers, tippers
- (B3e) Dumpers, dump trucks, skip dumpers
- (B3f) Passenger cars and buses
- (B3g) Wheel tractors, crawler tractors and winches, including battery operated electric tractors if not in (B2f)
- (B3h) Platform, fork-lift and straddle trucks
- (B3i) Wheelbarrows and wagons, concrete prams, wheel bogeys, piano trolleys
- (B3j) Conveyors, roller conveyors, belt conveyors if not in (B2k)
- (B3k) Locomotives, wagons, railroad equipment and track, monorail equipment
- (B3m) Other haulage plant
- (B3p) Lifting and hoisting plant, (mainly vertical)
- (B3q) Mobile jib cranes, eg: wheeled, truck or crawler mounted cranes
- (B3r) Derricks
- (B3s) Tower derricks
- (B3t) Tower cranes, eg: rail mounted, static or climbing cranes, pneumatic-tyred, truck or crawler mounted tower cranes
- (B3u) Bridge type cranes
- (B3v) Other types of cranes
- (B3w) Lifts, hoists
- (B3x) Grabs, skips, slings, winches, pulley blocks and gin wheels, and lifting gear in general
- (B3y) Other transport plant

(B4) Manufacture, screening, storage plant
Hand tools if described separately see (B7)
- (B4d) Crushers, screens, feeders, sieves
- (B4e) Mastic asphalt plant
- (B4f) Rolled asphalt and coated macadam plant
- (B4g) Concrete batching, mixing and placing plant
- (B4h) Pre-stressing equipment
- (B4i) Mortar machines
- (B4k) Hoppers, tanks, bulk cement handling and storage plant
- (B4m) Other manufacturing plant

(B5) Treatment plant
Drilling, cutting, welding, grinding, painting etc
Hand tools if described separately see (B7)
- (B5c) Earth and rock drilling equipment
- (B5d) Riveting, screwing and nailing tools, chippers
- (B5e) Cutting machines and welding equipment
  arc welding, thermic boring etc
- (B5f) Grinding and drilling machines
- (B5g) Cutting and bending machines
- (B5h) Blasting, puttying and spray-painting equipment
- (B5j) Workshop equipment
  benches, lathes etc
- (B5m) Other treatment plant

(B6) Placing, pavement, compaction plant
Placing, fixing;
Hand tools if described separately see (B7)
- (B6b) Pavement equipment in general
- (B6c) Asphalt and coated macadam spreaders
- (B6d) Concrete paving machines
- (B6e) Compaction equipment in general
- (B6f) Compaction equipment for earth works
  dead weight rollers, vibrating rollers, vibrating plates, power rammers etc
- (B6g) Compaction equipment for concrete
  immersion vibrators, clamp-on vibrators, surface vibrators etc
- (B6h) Soil stabilization equipment
- (B6j) Plastering and gun machines
- (B6m) Other placing, pavement and compaction plant

See Appendix 6

| | (B) Construction plant, tools* |
| | (D) Construction operations* |

(B)/(D)

(B7) Hand tools
Hand held tools
Hand operated (non-powered) tools

(B8) Ancillary plant
Plant which is not used directly for building work
(B8b) Power, heating and ventilation
(B8d) Motors
(B8e) Generators, transformers, distribution boxes and electric material
(B8f) Air compressors
(E8h) Steam boilers and steam generators
(E8i) Drying and heating equipment
(B8j) Refrigeration plant
(B8k) Fans
(B8m) Other equipment
(B8p) Other equipment for building, landscaping and civil engineering works
Drawing, surveying equipment see (A3)
(B3q) Investigating equipment
(B3r) Signalling equipment
(B3u) Cleaning equipment
(B8v) Testing equipment
See also (B2q)
(B8w) Horticultural sundries eg: ties, stakes
(B8x) Landscape maintenance equipment

(B9) Other construction plant, tools

(C) Vacant*
Code (C) may be used for 'Labour only' Activities, Trades, Personnel eg: bricklayers, joiners

(D) CONSTRUCTION OPERATIONS*
If necessary subdivide eg:

(D1) Protecting
All operations aimed at safeguarding and protecting the job
Plant, equipment for protecting see (B1)

(D2) Clearing, preparing
Initial preparation for the job
Plant for preparation see (B2)
Surveying see (A3)
casting, demolishing, clearing and levelling, setting out, excavating, trenching, strutting, timbering, installing temporary services

(D3) Transporting, lifting
Handling materials to be used for the job
Plant for transport see (B3)
carrying, handling, hoisting, lifting, loading, pumping, unloading

(D4) Forming: cutting, shaping, fitting
Changing the basic shape and size of material eg:
operations resulting in waste
cutting, grinding, sand-blasting
other forming operations
bending, dressing, texturing, smoothing, mixing, twisting, lapping, mowing, pruning, riveting, welding, rolling, flattening

(D5) Treatment: drilling, boring
Working on materials without changing their basic size:
boring, drilling, morticing

(D6) Placing: laying, applying
Locate and fix operations:
assembling, coating, pouring, compacting, fixing, installing, lagging, laying, placing, securing, tamping, driving

(D7) Making good, repairing
Completion of (D4), (D5) and (D6), clearing up, completing,
Alterations, repairs in general see (W)

(D8) Cleaning up

(D9) Other construction operations

* See Appendix 6

77

(E)/(G)

- (E) Composition etc
- (F) Shape, size etc
- (G) Appearance etc

Requirements, Properties, Processes (E)/(Y)
'Building science', construction technology

(E) to (G) are factors describing things such as buildings, elements, materials etc, and (H) describes their environment;
(J) to (T) are performance factors, factors relating to use;
(U) types of users and their requirements as individuals and groups;
(V) to (W) are concerned with working, operation and maintenance factors;
(X) with change factors; and
(Y) with commercial factors.

(E) to (Y) may be used to subdivide (A) to (D) and vice versa eg:
(B2d) (J3) Strength factors for scaffolding
(W) (A1) Organising for maintenance

Description (E)/(G)

(E) COMPOSITION ETC
If necessary subdivide eg:

(E1) Content as a whole
constituents, parts, finishes
Factors relating to internal structure, state, mix proportions
Factors relating to fluids, mixtures and solids see (L)

(E2) Production factors
(E3) Manufacture, prefabrication methods, factors
(E4) Assembly ie on-site assembly methods, factors
including building construction
(E5) Other production factors eg:
productivity, industrialization, automation

(E7) Factors relating to connection and fixing
Fixings see t

(E8) Factors relating to accessories

(E9) Other factors relating to composition, content

(F) SHAPE, SIZE ETC
If necessary, subdivide eg:

(F1) Shape, geometry eg:
spatial shape, relationships; layout, siting plan or shape

(F4) Size
3 Dimensional co-ordination, modular co-ordination
5 Volume
  cubic content, capacity
6 Area
7 Dimensions eg:
  length, span, distance; height, altitude, elevation; depth; width (breadth); perimeter; radius; diameter including bore (internal diameter); thickness; slenderness; scale (relative dimension);
  Dimensional change eg: shrinkage, expansion, movement

(F5) Weight, density

(F6) Tolerance; fit; accuracy

(F7) Dimensional systems eg:
metric system

(F8) Other factors relating to layout, shape, dimensions etc

(G) APPEARANCE ETC
Sensory factors as a whole
If necessary, subdivide eg:

(G1) Appearance; visibility; composition, scale (aesthetic); form, proportion; style

(G4) Texture eg:
roughness, smoothness; flatness (evenness); pattern

(G5) Colour eg:
value; chroma; chromatic colour including hue; chromaticism; iridescence; lustre

(G6) Opacity
(translucency, transparency)

(G7) Other appearance (visual) factors

(G8) Sensory factors other than visual eg:
feel (tactile) including warmth to touch; smell including scented, odourless; aural including audibility; taste

(G9) Other descriptive factors

(H) **Context, environment**
(J) **Mechanics**
(K) **Fire, explosion**

(H)/(K)

(H) CONTEXT, ENVIRONMENT
Factors relating to surroundings, occupancy
If necessary, subdivide eg:

(H1) Natural environmental factors
1 Regional factors eg:
   polar, temperate, sub-tropic, tropic
2 Meteorological factors eg:
21 Climate (weather) including wet, dry, hot and cold weather; extremes of temperature; microclimate
22 Parts of weather eg: sunshine; air currents, winds including gales, cyclones (hurricanes); storms including lightning; fog; frost; rain; snow; hail; ice
23 Weather incidence, exposure; weather tightness; weather proofing, protection, resistance
28 Other meteorological factors
4 Weathering
5 Topographical factors
6 Physiographic, geological factors eg:
   landslides, avalanches; tides; droughts; floods; volcanic eruptions; earthquakes; subsidence
7 Ecological factors eg:
   'clean air', 'clean water'
8 Other natural environmental factors

(H2) Factors relating to modified environments eg:
overshadowing, wind effects around tall buildings, traffic, smog

(H4) Amenities

(H6) Factors relating to internal environments eg:
draughts

(H7) Occupancy factors eg:
types of usage including industrial, commercial, recreational, educational, religious, residential
Other occupancy factors eg overcrowding

(H8) Other factors relating to surroundings

Performance factors (J)/(T)

(J) MECHANICS
Statics and dynamics factors
If necessary, subdivide eg:

(J1) Mechanical factors for solids
1 Mechanics of structures, statics
2 Soil mechanics, geotechnics, subsidence, settlement

(J2) Mechanical factors for fluids
1 Fluid mechanics, hydraulics
2 Gas mechanics, aerodynamics

(J3) Strength, resistance to deformation

(J4) Loads, stresses, forces, eg:
compression, tension, bending; shear; torsion; impact strength; bursting; tearing; splitting; cracking; holding power; hardness (indentation); wear (abrasion).
Other types of loads eg dead loads, live loads, wind loads.

(J5) Forces opposing motion eg:
adhesion, friction (including slipperiness, viscosity)

(J6) Vibration factors eg:
sources, protection

(J7) Deformation factors eg:
flexibility, elasticity, plasticity, yield point;
stability, failure, collapse

(J8) Other factors relating to statics, dynamics, etc

(K) FIRE, EXPLOSION
If necessary, subdivide eg:

(K1) Fire types, sources

(K2) Fire protection
1 Retardant treatment
2 Spread prevention including fire stopping
3 Control including fire fighting, venting
4 Means of escape

(K3) Fire resistance (structures)

(K4) Reaction to fire (materials)
1 Combustibility including combustible, non-combustible (incombustible)
2 Calorific value (thermal value)
3 Fire load, heat release rate
4 Ignitability including ignition temperature (fire point), flash point
5 Flammability (inflammability), flame spread

(K5) Combustion products eg:
smoke, fumes, soot

(K6) Explosions eg:
arresting, venting; explosive limits (gases and vapours)

(K7) Fire and explosion damage, salvage
Damage see (X7)
Salvage see (X8)

(K8) Other factors relating to fire, explosion

(L)/(M)

(L) Matter
(M) Heat, cold

(L) MATTER
Factors relating to interactions with matter (fluids, solids)
Factors relating to content as a whole see (E1)
If necessary, subdivide eg:

(L1) Fluids

(L2) Factors relating to gases, vapours
including air, water vapour (ie interaction properties of material etc with gases, vapours), ventilation
1 Humidity including absolute, relative
3 Gas absorption including (for water vapour) hygroscopy deliquescence
4 Gas diffusability
6 Gas permeability; gas tightness; gas proofing
7 Condensation including interstitial condensation
8 Other factors related to gases

(L3) Factors relating to liquids and mixtures (mixes)
including water, solutions (with liquids, mixtures)
1 Liquid content (moisture content, dampness, dryness) including rising damp
3 Liquid absorption
4 Liquid permeability; liquid tightness including water tightness; liquid proofing including waterproofing; damp proofing (moisture proofing)
5 Capillarity
6 Drying
7 Dissolving including solvency, solubility, miscibility
8 Other factors relating to liquids, mixtures

(L4) Factors relating to solids
1 Hydration
2 Penetration (solid permeability)
4 Deposition eg: efflorescence, pattern staining; corrosion, erosion
8 Other factors relating to solids

(L5) Physical and chemical factors related to matter
1 Physical factors eg: permeability, viscosity
2 Chemical factors eg: reactivity, and processes such as hydrolysis, reduction, base exchange, impurity
Protection, proofing against chemical change

(L6) Biological change factors
Rot, decay, interaction with biological entities
1 Human, such as accidental damage, vandalism
2 Animals, birds, insects (including vermin), such as infestation
3 Microbes (including bacteria), such as disease
4 Plants (including fungi, mould), such as dry rot
Protection, proofing against biological change

(L7) Contamination factors
Pollution (by effluents, dust etc)
Purification, protection, proofing against contamination

(L8) Other factors relating to matter eg:
deterioration and protection, proofing against deterioration

Heat, Light, Sound, Electricity (M)/(Q)
Include here information relevant to Heat + Light + Sound etc eg:
insulation as a whole, protection against two or more of these energy forms

(M) HEAT, COLD
Thermal factors
If necessary, subdivide eg:

(M1) Heat types, sources

(M2) Thermal insulation
Heat/cold protection, proofing

(M3) Thermal transfer
1 Thermal transmission eg: transmittance including U value (total thermal resistance); transmissivity including surface coefficient; thermal resistance including resistivity; surface resistance
3 Conduction including conductance, conductivity
4 Convection including convection coefficient
5 Radiation including radiation coefficient
6 Thermal absorption
7 Thermal emission
8 Other factors relating to thermal transfer

(M4) Thermal capacity (heat capacity), specific heat

(M5) Heat loss, gain eg:
solar heat gain, insolation including thermal expansion, thermal contraction if not at (F47)

(M6) Overheating, inadequate heat

(M)	Heat, cold	
(N)	Light, dark	
(P)	Sound, quiet	**(M)/(P)**

(M7) Temperature eg:
surface temperature
absolute including Kelvin temperature; Celsius (centigrade) temperature; Fahrenheit temperature; cryogenic (low temperature), refractory (high temperature) factors

(M8) Other factors; relating to heat, cold eg:
control; detection; thermal shock

(N) **LIGHT, DARK**
Optical factors
If necessary, subdivide eg:

(N1) Light types, sources
1 Natural light including daylight, sunlight
2 Artificial light
3 Combined natural and artificial light including PSALI

(N2) Light insulation
Light protection, proofing; light tightness

(N3) Light transmission
including diffuse transmission
Light absorption; emission
including incandescence, luminescence

(N4) Light reflection; light polarisation; light refraction
1 Light reflection eg:
reflection factor (reflectance, reflectivity) including diffuse reflection factor, direct reflection factor; reflecting including mirror; non-reflecting including matt
2 Light polarisation
3 Light refraction including refractive index

(N5) Brightness (luminosity, brilliance)
Light (bright), dark, luminance

(N6) Glare
including glare index

(N7) Light flux
Illumination including sky factor, daylight factor
Candlepower (luminous intensity)

(N8) Other factors relating to light eg:
control; detection

(P) **SOUND, QUIET**
Acoustic factors
If necessary, subdivide eg:

(P1) Noise types, sources
electro acoustics

(P2) Sound insulation
Sound protection; sound proofing; sound tightness

(P3) Sound transmission
including airborne sound transmission, impact sound transmission; sound transmission factor;
Sound resistance
including resonance
Sound absorption
including reverberation

(P4) Sound reflection
including echo

(P5) Sound gain, loss

(P6) Loudness, quietness

(P7) Measurement of sound
hertz, decibels

(P8) Other factors relating to sound eg:
control; detection

(Q)/(T)

(Q)	Electricity, magnetism, radiation
(R)	Energy, other physical factors
(T)	Application

(Q) ELECTRICITY, MAGNETISM, RADIATION
If necessary, subdivide eg:

(Q1) Electricity types, sources
 1 Current electricity factors including short circuits;
 2 Static electricity factors

(Q2) Electrical insulation
including dielectric constant
Electrical protection, screening

(Q3) Electrical conduction
including conductance, conductivity
Electrical resistance
including admittance

(Q4) Electrical ionisation
including ionisation potential

(Q5) Voltage (electromotive force)
Capacitance

(Q6) Factors relating to magnetism and electromagnetism

(Q7) Factors relating to electromagnetic radiation
including solar, cosmic, nuclear (radioactivity, irradiation) ultraviolet, X-rays, infra red, micro wave
Light (visible radiation) see (N)
Protection, screening against electromagnetic radiation

(Q8) Other factors relating to electricity, magnetism, radiation eg:
control; detection
Light (visible radiation) see (N)

(R) ENERGY, OTHER PHYSICAL FACTORS
Including Side-effects, Compatability, Durability
If necessary, subdivide eg:

Energy (R1)/(R5)

(R1) Kinetic energy eg:
mechanical energy; fluids energy including hydromechanical (hydraulic), pneumatic and aerodynamic energy; speed; momentum; inertia; flow factors

(R2) Potential energy
including strain energy

(R3) Energy demand
including power requirement
Energy input, inflow
including fuel consumption; energy consumption (load, peak load, overload)
Energy output, outflow
including power capacity
Efficiency of energy conversion

(R4) Energy control, detection

(R5) Other factors relating to energy eg:
insulation, transmission, conservation

Other physical factors (R6)/(R8)

(R6) Side-effects

(R7) Compatibility

(R8) Durability

(S) Vacant

(T) APPLICATION
Including resources and their proper selection
If necessary, subdivide eg:

(T1) Factors relating to application for specific activities eg:
those relating to facilities in Table 0 classes 2/9 below eg:
(T1 93) Cooking

2	Industrial activities:
26	Agricultural activities eg: farming
27/28	Manufacturing activities eg: assembling cars
3	Administrative, commercial, protective activities:
31/32	Administrative activities eg: levying taxes
32	Business activities eg: office work
33/34	Commercial activities
33	Financing eg: accounting
34	Trading eg: buying
37	Protective service activities eg: fire fighting
4	Health, welfare activities:
41/42	Diagnostic, treatment activities eg: inspecting
44/46	Welfare activities eg: caring
5	Recreational activities:
51/53	Entertainment (social) activities eg: acting
54/56	Sporting activities eg: swimming
58	Play activities eg: gambling
6	Worshipping activities
7	Learning, information activities:
73/74	Research activities eg: experimenting
75	Display activities eg: exhibiting
76	Information activities eg: data processing
78	Teaching activities eg: teaching
8	Residential activities eg: living

		(T)	Application	**(T)/(U)**
		(U)	Users, resources	

9	Common activities	(U)	**USERS, RESOURCES**
91	Circulating activities (movement cf 11/14 large scale movement) Assembly activities eg: spectating		If necessary, subdivide eg:
92	Resting activities eg: sleeping Working activities eg: thinking	(U1)	**User groups, communities eg:** Hereditary groups including families, tribes, races Organised groups including households, societies, associations, nations Unorganized groups including crowds, masses, herds
93	Culinary, eating activities eg: cooking		
94	Sanitary, hygiene activities eg: personal washing		
95	Cleaning, maintenance activities eg: washing		
96	Storage activities		
97	Processing, controlling activities eg: supplying	(U2)	**Society, sociological factors eg:** Demography including population Customs, ways of life, behaviour patterns, morals ethics including peace, friendship, tolerance, alienation, hostility, aggression, violence Social economics including consumption, wealth, poverty, ownership, tenancy Social psychology including group consciousness, public opinion, convention, aspiration
99	Outdoor activities		

(T2) **Proper use, limitations on use**

(T3) **Suitability, efficiency, effectiveness**

(T4) **Usefulness, obsolescence, degree of utilization**

(T5) **Re-use; change in use, adaptability, flexibility**

(T6) **Consumption, waste, conservation**

(T7) **Mis-use, wrong use, mistakes in use**

(T8) **Failure, deficiency in use, defects**
Other aspects of use

(U3) **People (persons, human beings) including users**
1 Children
  Babies (infants)
  Adolescents (youths, minors, teenagers)
2 Adults including parents, fathers, mothers
  Old people (elderly people, aged people)
3 Females, women
  Males, men
4 People according to marital status eg: single (unmarried), married
5 Sick people (invalids) eg: addicts
  Handicapped people
  Physically handicapped eg: disabled (crippled), blind, deaf
  Mentally handicapped eg: retarded
6 People according to economic status eg: rich (wealthy, high income), low income, poor (needy, destitute); landlords, tenants, occupiers, homeless; employers, employees
7 People according to function, work, occupation eg: manual, administrative, professional; passengers; visitors; vandals
8 Other people, users

(U4) **Physical and mental factors**
1 Body factors
  Anatomical and physiological factors eg: nervous system, respiratory system
  Anthropometrics, ergonomics
  Other body factors eg: posture
5 Comfort, convenience
6 Health and hygiene factors including related preventative, abatement and control factors
  1 Types of illness
  3 Nutrition factors
  4 Cleanliness factors
  5 Sanitation factors
  6 Disease, infection factors including toxicity; bacteria, viruses, parasites
  7 Stress factors
7 Safety and security factors, risks, hazards, accident prevention
  Fire, explosion see (K)
8 Other physical and mental factors eg: response to the environment, aesthetic appreciation

(U5) **Non-human 'users' and requirements**
1 Animals including insects, fishes, reptiles, birds, mammals
  Domestic, farm animals
  Wild animals including game, wild fowl
  Factors relating to animal husbandry and animal produce
2 Plants
  Factors relating to plant husbandry and plant products
3 Machines
  Factors relating to machines

(U6) **Resources eg:**
natural, man-made; manufacturing, management; energy; space

(U8) **Other factors relating to users, resources eg:**
those relating to life

(V)/(X)

- (V) Working factors
- (W) Operation, maintenance factors
- (X) Change, movement, stability factors

(V) WORKING FACTORS
Factors relating to workability
If necessary, subdivide eg:

(V1) Storage life, shelf life
(ease of storing)

(V2) Demountability, detachability etc
(ease of dismantling)

(V3) Movability, manoeuvrability etc
(ease of moving)

(V4) Ease of cutting

(V5) Ease of drilling

(V6) Ease of placing, connecting
consistency, spreadability, hiding power, drying etc (ease of coating)
weldability, adhesiveness, cohesiveness etc (ease of joining)
nailability etc (ease of fixing)

(V7) Adaptability, adjustability, interchangeability
ability to accept other units and accessories etc (ease of altering)

(V8) Other working factors eg:
ease of cleaning, disposability

(W) OPERATION, MAINTENANCE FACTORS
If necessary, subdivide eg:

(W1) Method of operation

(W2) Maintenance, routine servicing and cleaning
including re-decorating

(W5) Overhaul and repair

(W6) Alteration, adaptation, modification, conversion

(W7) Restoration, reconstruction, renovation, renewal, replacement
improvement, modernisation, rehabilitation

(W8) Other factors relating to operation and maintenance

(X) CHANGE, MOVEMENT, STABILITY FACTORS
If necessary, subdivide eg:

(X1) Associative change
including integration, acquisition, concentration, addition, absorption

(X2) Dissociative change
including disintegration, separation, differentiation, subtraction, leakage, emission, distribution, removal, extraction

(X3) Transfer
Movement, rhythm

(X4) Considerable or complete change
including disturbance, transformation, renewal, substitution, revolutionary change

(X5) Little or no change,
including stability (equilibrium, balance)

(X6) Gradual change
including evolution (development, growth), devolution, decline, ageing, maturing, curing

(X7) Quality change
Improvement eg:
modernisation, reinforcement, toughening, stiffening, purification
Deterioration eg:
dilapidation, dereliction, decay, wear, damage, breakage, weakening

(X8) Causes, effects of change
Introduction (realisation, execution) of change eg:
innovation; discarding (scrapping); reclamation (recovery, salvage); restoration
Control of change
interaction of, resistance to, prevention of change
Protection from, insulation from, proofing from change
Preservation, conservation

(Y)	Economic, commercial factors	
(Z)	Peripheral subjects, form of presentation, time, place	**(Y)/(Z)**

(Y) ECONOMIC, COMMERCIAL FACTORS
If necessary, subdivide eg:

(Y1) Economic factors
Availability of finance

(Y2) Costs, prices
(Y3) Initial, capital costs
including purchase price, conditions including guarantees, method of payment; construction price, conditions including guarantee, method of payment
(Y4) In-use, running costs

(Y6) Values, benefits

(Y7) Economic efficiency
Other economic factors

(Y8) Supply factors
Sources of supply
Availability
Supply capacity, programmes, delivery period
Packaging, labelling
Ordering
Delivery
Other supply factors

(Y9) Services factors
Contract services eg: hire services
Advisory services
Design and Construction services eg: package deal

(Z) PERIPHERAL SUBJECTS, FORM OF PRESENTATION, TIME, PLACE

This position may be used to introduce an arrangement of subjects peripheral to the construction industry, taken from UDC. The UDC number is added inside the bracket eg: (Z31) Statistics.

UDC Common Subdivisions for form of presentation, time and place are also introduced with (Z) eg: (Z(03)) Dictionaries; (Z '19') 20th Century; (Z(42)) England. They may be used with CI/SfB or with UDC peripheral classes.

Outline schedules of UDC main classes and UDC based common subdivisions for form of presentation, time and place are given on the following pages.

Forms of presentation file before time and time files before place.

It is not necessary to use the symbol Z more than once eg: (Z31(03)) Dictionary of Statistics.

Main classes coded by UDC may be preceded by the letters UDC.

85

Main classes of UDC*

	Generalities
	Prolegomena. Fundamentals of knowledge and culture
01	Bibliography. Catalogues
02	Libraries. Librarianship
03	Encyclopaedias. Dictionaries. Reference books
04	Essays. Pamphlets, offprints, brochures and the like
05	Periodicals. Reviews
06	Corporate bodies. Institutions. Associations. Congresses. Exhibitions. Museums
07	Newspapers. Journalism
08	Polygraphies. Collective works
09	Manuscripts. Rare and remarkable works. Curiosa

1	**Philosophy; Metaphysics; Psychology; Logic; Ethics and morals**
12	Metaphysics
13	Metaphysics of spiritual life. Occultism
14	Philosophical systems
15	Psychology
16	Logic. Theory of knowledge. Logical method
17	Ethics. Moral science. Convention
18	Aesthetics

2	**Religion; Theology**
21	Natural theology
22	Holy Scripture. The Bible
23	Dogmatic theology
24	The religious life. Practical theology
25	Pastoral theology
26	The Christian church in general
27	General history of the Christian church
28	Christian churches or worshipping bodies
29	Non-Christian religions. Comparative religion

3	**Social sciences; Economics; Law; Government; Education**
30	General sociology. Sociography
31	Statistics
32	Political science. Politics. Current affairs
33	Political and social economy. Economics
34	Jurisprudence. Law. Legislation
35	Public administration. Military science. Defence
36	Social relief and welfare. Insurance
37	Education
38	Trade. Commerce. Communication and transport
39	Ethnography. Custom and tradition. Folklore

4	**Philology; Linguistics; Languages**
41	Philology and linguistics in general
42	Western languages in general. English
43	Germanic languages. German. Dutch, etc
44	Romance languages in general. French
45	Italian. Roumanian, etc
46	Spanish. Portuguese, etc
47	Latin and Greek
48	Slavonic languages. Baltic languages
49	Oriental, African and other languages

5	**Mathematics and natural sciences**
51	Mathematics
52	Astronomy. Surveying. Geodesy
53	Physics and mechanics
54	Chemistry. Crystallography. Mineralogy
55	Geology. Meteorology
56	Palaeontology
57	Biology. Anthropology
58	Botany
59	Zoology

6	**Applied sciences; Medicine; Technology**
61	Medical sciences. Health and safety
62	Engineering and technology generally
63	Agriculture. Forestry. Stockbreeding. Fisheries
64	Domestic science. Household economy
65	Commercial, office, business techniques. Management. Communications. Transport
66	Chemical industry. Chemical technology
67	Manufactures, industries and crafts
68	Specialized trades, crafts and industries
69	Building industry, materials, trades, construction

7	**The arts; Recreation; Sport; etc.**
71	Physical planning. Landscape, etc
72	Architecture
73	Sculpture and the plastic arts
74	Drawing and minor arts and crafts
75	Painting
76	Engraving and prints
77	Photography. Cinematography, etc
78	Music
79	Entertainment. Pastimes. Games. Sport

8	**Literature; Belles lettres**
80	Generalities. Rhetoric. Criticism
82	Literature of the Western countries. English literature
83	Germanic literature: German, Dutch and Scandinavian
84	Romance literature. French literature
85	Italian literature. Roumanian literature
86	Spanish literature. Portuguese literature
87	Classical, Latin and Greek literature
88	Slavonic literature. Baltic literature
89	Oriental, African and other literature

9	**Geography; Biography; History**
91	Geography, exploration and travel
92	Biography. Genealogy. Heraldry
93	History in general. Sources. Ancient history
94	Mediaeval and modern history
940	History of Europe
950	History of Asia
960	History of Africa
970	History of North America
980	History of South America
990	History of Oceania, Australasia and Polar regions

*From BS 1000A:1961 Universal Decimal Classification, Abridged English edition (3rd edition, London, BS1 1961) and reproduced by permission of the British Standards Institution, 2 Park Street, London W1A 2BS.

Common auxiliaries of UDC

FORM OF PRESENTATION
from table (d) in BS 1000A:1961

- (02) Books, textbooks, manuals
- (03) Dictionaries, encyclopaedias, glossaries
- (04) Brochures, addresses, theses, letters, articles, reports
- (05) Journals, magazines, yearbooks
- (06) Publications of particular societies, institutions
- (07) Instruction books, study books
- (08) Tables, graphic representations
- (09) Historical, biographical, legal

TIME
from table (g) in BS1000A:1961

Centuries or decades are denoted by 2 or 3 digit notations respectively eg:
- '03' 4th century (300's)
- '19' 20th century (1900's)
- '193' the thirties (1930-39)

Periods embracing several centuries, etc thus:
- '03/08' 4th to 9th century
- '1919/1939' the 'inter-war' years
- '187/189' 1870 to 1890
- '−' Antiquity BC
- '+' Christian Era AD

PLACE
from table (e) in BS1000A:1961

- (1) Place in general
- (2) Physiographic designation
- (3) The ancient world

The modern world (4/9)

- (4) Europe
- (41/42) British Isles (geographical whole)
- (41-4) United Kingdom of Gt. Britain and N. Ireland
- (41-44) Commonwealth
- (410) Great Britain (geographical whole)
- (411) Scotland
- (415) Ireland (geographical whole)
- (416) Northern Ireland
- (417) Irish Republic (Eire)
- (420) England
- (429) Wales
- (430) Germany
- (436) Austria
- (437) Czechoslovakia
- (438) Poland
- (439) Hungary
- (44) France
- (45) Italy
- (46) Spain
- (469) Portugal
- (47) USSR
- (48) Scandinavia
- (480) Finland
- (481) Norway
- (485) Sweden
- (489) Denmark
- (492) Netherlands
- (493) Belgium
- (494) Switzerland
- (495) Greece
- (497) Yugoslavia, Bulgaria

- (5) Asia
- (51) China, Korea etc
- (52) Japan
- (53) Arabia
- (54) India, Ceylon, Pakistan
- (55) Iran (Persia)
- (56) South West Asia Turkey, Cyprus, Iraq, Syria, Lebanon, Israel, Jordan
- (57) USSR Far East
- (58) Central Asia eg: Afghanistan
- (59) South East Asia Burma, Thailand, Malaysia, Cambodia, Vietnam, Laos

- (6) Africa
- (61) Tunisia, Libya
- (62) Egypt, UAR, Sudan
- (63) Ethiopia
- (64) Morocco
- (65) Algeria
- (66) W & N Central Africa eg: Ghana, Nigeria
- (67) S Central, E Africa eg: Congo, Uganda, Kenya, Somaliland, Tanzania, Mozambique
- (68) Southern Africa eg: South Africa, Rhodesia
- (69) Islands around Africa

- (7) North America
- (71) Canada
- (72) Mexico
- (728) Central America
- (729) West Indies
- (73) USA
- (74) N E States
- (75) S E States
- (76) S Central States
- (77) N Central States
- (78) W States
- (79) Pacific States

- (8) South America
- (81) Brazil
- (82) Argentina
- (83) Chile
- (84) Bolivia
- (85) Peru
- (86) Columbia
- (866) Ecuador
- (87) Venezuela
- (88) Guyana
- (892) Paraguay
- (899) Uruguay

- (9) Oceania, Arctic and Antarctic Regions
- (91) Malay Archipelago East Indies
- (910) Indonesia
- (914) Philippines (Republic)
- (93) Australasia generally, Melanesia
- (931) New Zealand
- (94) Australia
- (95) New Guinea (and Papua)
- (96) Polynesia and Micronesia Micronasia
- (98/99) Polar Regions
- (98) Arctic and N Polar Regions
- (99) Antarctica South Polar Regions

Index
to the tables

Introduction

Capital letter entries refer to the main headings in the tables, which can be regarded as CI/SfB key terms. They are terms with one and two digit codes in Tables 0, 1 and 4 (classes (A)/(D)) eg:

ROAD TRANSPORT FACILITIES	12
LIQUIDS SUPPLY SERVICES	(53)
PROJECT ORGANISATION	(A1)

and single digit codes in Tables 2, 3 and 4 (classes (E)/(Y)) eg:

SECTIONS AND SECTION WORK	H
METAL	h
FIRE	(K)

These CI/SfB main headings are given on the CI/SfB wall chart.

Italic entries in the index are derived from Appendix 1. The whole of this appendix can be included in Table 1 at (99); but wherever possible it is preferable to relate Appendix 1 terms either to particular elements or to main element groups eg:

Reinforcement	*(99.79)*
to building frame	*(28.9)*
to floors	*(23.9)*
to roofs	*(27.9)*
to structure generally	*(29)*
to walls	*(21.9)*

– use codes such as *(23.9)* and *(29)* rather than *(99.79)*.

Such index entries are not exhaustive statements of elements to which individual Appendix 1 terms apply (reinforcement is relevant to elements other than those listed). They illustrate how the figure 9 representing any of the terms in Appendix 1, can be used as a subdivision of element codes to indicate the idea of 'parts'. To avoid giving undue prominence to Appendix 1 (Table 1 concepts are far more important), only its main headings appear in the index.

To use the index, look first for the term which most specifically describes the sought item, and only look for more general terms if this is not included. Always check the context of items found in the index against the schedules. If the codes are longer than you require for your particular application of CI/SfB, simply remove digits from the right. In the example above, *reinforcement to floors* could be represented as *(23)* and *reinforcement* as *(99)*.

Information on the construction of simple indexes and on post-coordinate indexing is given in Appendix 3.

Aba–Ant

A

Abattoirs 273
Abbeys 66
Ablutions facilities 942
Above ground drainage (52.6)
Abrasion (J4)
Abrasives w7
Absolute humidity (L21)
Absolute temperature (M7)
Absorbers, ultra-violet u9
Absorbing glass, heat o6
Absorption (X1)
    gas (L23)
    light (N3)
    liquid (L33)
    sound (P3)
    thermal (M36)
Abutment retaining walls (16.2)
Academies 724
Accelerators u2
Access facilities 913
Access fittings (71.3)
    loose furniture and equipment (81)
Access floors (33.1)
Access roads 1224
Access traps (roof secondary elements) (37.5)
Accesses (99.6)
Accessories (99.99)
    characteristics of (E8)
Accident prevention (U47)
Accidental damage (L61)
Accommodation (administration)
    costs (A2t)
    project (or office) administration (A1r)
Accommodation (site facilities) (B1b)
ACCOUNTING (A2)
Accounts (A2r)
Accuracy (F6)
Acetal resins n6
Acetate rayons n6
Acids w5
ACOUSTIC FACTORS (P)
Acquisition (X1)
Acrobatic facilities 562
Acrylic resins n6
Action areas 0832
ACTIVITIES
    in construction industry (A /(D)
    use of facilities (T1)
Activity data method (A3r)
Acts of Parliament (Ajk)
Adaptability
    application factors (T5)
    working factors (V7)
Adaptation (W6)
Addicts (U35)
Addicts welfare facilities 443
Addition (X1)
Adhesion forces (J5)
Adhesiveness (V6)
Adhesives t3
Adjustability (V7)
Administration
    activities (T131)/(T132)
    in construction industry (A)
Administrative areas 097
Administrative facilities 31/32
Administrative staff
    office organisation (A1my)
    users (U37)
Admission units (hospitals) 418
Admittance (Q3)
Admixtures u2
Adobe g1

Adolescents (U31)
Adult education facilities 727
Adults (U32)
Advance ordering (A6r)
Adventure playgrounds 585
Advertising (Ai)
    personnel selection (A1mq)
    project (or office) administration (A1i)
    staff selection (A1mn)
Advice
    official (Ajp)
    technical services (Y9)
Advisory leaflets, DOE (Aju)
Advisory publications (Ajp)
Aerated concrete q6
    precast f4
Aerial masts 156
Aerial ropeways 113
Aerials (99.125)
    communications services in general (64.9)
    radio services (64.31)
    television services (64.11)
Aerodromes 141
Aerodynamic energy (R1)
Aerodynamics (J22)
Aeronautical sports facilities 567
Aesthetics (G1)
    physical responses (U48)
Aged people
    see Old people
Ageing (X6)
AGGREGATES p
Aggression (U2)
Agreements
    contracts (A4u)
    personnel selection (A1mq)
Agrément certificates (Ajs)
Agricultural activities (T126)
AGRICULTURAL FACILITIES 26
Agricultural wheeled tractors (B2f)
AIDS (A)/(D)
Aikido facilities 562
Air (L2)
    clean (H17)
    currents (H122)
Air compressors (B8f)
AIR CONDITIONING AND VENTILATION SERVICES (57)
Air conditioning services (57.1)/(57.5)
    central air conditioning (57.1)
    unit air conditioning (direct, locally acting) (57.3)
    Code (57.1) is used as the general position
Air cushion transport 149
Air entrained concrete
    see Aerated concrete
Air fields 141
Air force facilities 3751
Air houses (site protection) (B1d)
Air raid shelters 3757
Air sports facilities 567
Air supply services (54.3)
Air supported structures (2.8)
Air traffic control towers 146
AIR TRANSPORT FACILITIES 14
Airborne sound transmission (P3)
Airbricks F
Air-cooled blast furnace slag p2
Aircraft 148
    control facilities 146
    storage and repair facilities 147
Air-entraining agents u2
Airing cupboards (75.3)

Airing facilities 953
Airport facilities 141
Aisles 68
Alabaster r2
Alarm glass o8
Alarms (99.12)
    fire protection services (68.52)
    intruder security services (68.2)
    site protection (B1d)
Alienation (U2)
Alkalis w5
Alleys 9141
All-in aggregate concrete q4
    precast f2
'All-in' service organizations (Aky)
Alloys
    aluminium h4
    copper h6
    other metals h9
    steel h3
Alpine plants W
Altars 68
Alteration (W6)
    ease of altering (V7)
Altitude (F47)
Alumina cement, high q2
Alumina-silica (fireclay) g6
Aluminium h4
    aluminium alloys h4
    aluminium bronze (copper alloy) h6
Aluminium oxide aggregate (for granolithic and terrazzo) q5
Ambulance facilities 373
Ambulatory facilities 68
Amenities (H4)
Amenity areas 097
Amorphous solids Y
Amplifiers (99.135)
    communications services (64.9)
Amusement arcades 582
Anatomical factors (U41)
Anchoring devices t6
    or use (99.94)
Ancient monuments 986
Ancillary materials w
Ancillary plant (B8)
Angle blocks F
Angle dozers (B2g)
Angle joints Z
Angles (sections) H
Anhydrite r2
Anhydrous gypsum plaster r2
Animal materials j
    glues t3
Animal welfare facilities 46
    see also Livestock facilities
Animals
    factors affecting works (L62)
    fauna 0874
    requirements (U51)
Annuals (plants) W
Anodised coatings u1
Antennae (99.125)
    communications services in general (64.9)
    radio services (64.31)
    television services (64.11)
Anthropometrics (U41)
Anti-corrosive materials u1

Consult the schedules to check the context of items in the index. Italic entries are explained in the introduction.

Ant–Beh

Anti-foaming agents u2
Anti-glare glass o7
Anti-settling agents u2
Anti-skinning agents u2
Anti-slip properties (J5)
Anti-static agents u9
Apartments 816
Appeals (Ajy)
    designing and physical planning (A3j)
APPEARANCE (G)
APPLICATION (T)
Applications for employment (A1mo)
Applied finishes (4-)
Applying (construction operations) (D6)
Appointments of personnel (A1mq)
APPRAISAL (A8)
Approach roads 1224
Approvals (designing and physical planning) (A3j)
Approved schools 7175
Aprons (air transport) 148
Aquaria 755
AQUATIC SPORTS FACILITIES 54
Aqueducts 1821
Arbitration (Aqr)
Arc welding equipment (B5e)
Arcades
    amusement 582
    shopping 342
Arcades (circulation facilities) 9141
Arch bridges 1828
Arched roofs (27.6)
Archery ranges 565
Arches (99.72)
Architects (Amw)
    buildings by particular architects 987
    in commerce (Aky)
    in private practice (Akv)
    in the public sector (Aks)
'Architectural ironmongery' t7
ARCHITECTURE AS A FINE ART 999
Architecture, schools of (Ano)
Architraves (31.9)
    for doorways (31.59)
    for window openings (31.49)
Archives 767
Area (F46)
Areas, planning 0
Areas of outstanding natural beauty 0873
Arenas
    entertainment 52
    sport 568
Armchairs (72.2)
    loose furniture and equipment (82)
Armed forces facilities 3751/3753
Arresting explosions (K6)
Art colleges 722
Art galleries 757
Art studios 32
    common facilities 926
Arterial roads 1221
Articles found on site (Aqt)
Artificial aggregates p2/p3
Artificial light (N12)
Artificial stone f3
Artists (Amw)
Arts centres 532
Asbestos based mixes q9
Asbestos cement f6
Asbestos wool m2
Asbestos-silica-cement f6
Asbestos-silica-lime f6
Ash p4
Ashlars F
Aspects (facades) 995

Asphalt s1
    mastic s4
    preformed n1
    rolled s5
Asphalt manufacturing plant (B4e)/(B4f)
Asphalt spreaders (B6c)
Aspiration (U2)
Assembled floors (23.4)
Assembly (circulation) (T191)
Assembly (construction operations) (D6)
    production factors (E4)
Assembly facilities 916/918
Assembly lines 285
Assize courts 317
Associations (U1)
    construction industry (Ak)
    trade (Akp)
Associative change (X1)
Athletics facilities 564
Atmospheric areas 091
    international or national in scale 02
Attachment devices t6
    or use (99.94)
Attenuators of noise/sound (99.535)
    air conditioning and ventilating services (57.9)
    sound control services (68.8)
Attics 993
Auction rooms 341
Audibility (G8)
Audience spaces 528
Audio communications services (64.3)
Audio-visual communication (Agi)
Audio-visual communications
    services (64.1)
    security services (68.2)
Audio-visual facilities 768
    cinemas 525
Auditing (A2j)
Auditoria 528
Aural factors (G8)
Autobahns 121
Automatic operating devices t7
    or use (99.964)
Automatically operated fire fighting services (68.54)
Automation factors (E5)
Availability (Y8)
Avalanche protection facilities 1861
Avalanches (H16)
Aviaries 753

B

BRE Digests (Ajt)
Babies (U31)
Back hoes (B2g)
Backing work
    thick coatings P
    wire netting J
Bacteria
    biological change factors (L63)
    health and hygiene factors (U466)
Bad debts (A2t)
Badminton facilities 562
Bains marie (culinary fittings) (73.5)
Balance (X5)
Balconies (23.7)
Balers (refuse disposal) (52.18)
Ball games facilities 56
Ball valves (99.411)
    liquids supply services (53.9)

Ballast p1
Ballrooms 521
Balustrades (99.51)
    roof secondary elements (37.6)
    secondary elements in general (38)
    stair and ramp secondary elements (34)
Bamboo j3
Bandstands 522
Banks 338
Banqueting rooms 518
Baptisteries 68
Bar charts (A5v)
Bar counters (catering fittings) (73.7)
Barges 138
Bark j5
Barns, dutch 2684
Barracks 3753
Barrages (water retention facilities) 187
Barred openings (99.5)
    in external walls (31.8)
    in internal walls (32.8)
    in floors (33.8)
Barrel lights (37.42)
Barriers (99.5)
    in external works (90.3)
Bars
    public houses 517
    snack bars 515
Bars (sections) H
Basalt e1
Bascule bridges 1825
Base exchange (L5)
Baseball facilities 564
Basement retaining walls (16.2)
Basements 993
Basins (washing fittings) (74.23)
Basket weave bond (blockwork and brickwork) F
Batching plant, concrete (B4g)
Bath houses 942
Bathroom suites (74.1)
Bathrooms 942
Baths (74.22)
    foot baths (74.24)
Battens H
Battered wives welfare facilities 445
Batteries (99.126)
    communications services (64.9)
    electrical services in general (69)
    electrical supply services (61.9)
Battery casting equipment (B2d)
Battery casting work E
Battery operated electric tractors (B3g)
    earth-moving plant (B2f)
Bays (building divisions) 994
Beaches 092
Beacons (boat control facilities) 136
Beaded joints Z
Beads (sections) H
Beam and column frames (28.2)
Beam and column structures (2.2)
Beam plus slab floors (23.4)
Beams (99.741)
    in building frames (28.9)
    in floors (23.9)
    in roofs (27.9)
    in structure generally (29)
Bedroom suites (72.1)
    loose furniture and equipment (82)
Bedrooms 921
Beds (72.1)
    loose furniture and fittings (82)
Beech i3
Bee-keeping facilities 4648
Beer gardens 517
Behaviour patterns (U2)

Bel–Buy

Belfries 68
Bells (99.125)
   audio communications (64.38)
   electrical services in general (69)
   fire detection and alarm (68.52)
   intruder detection and alarm (68.2)
   security and control services in general (68.9)
Below ground drainage (52.7)
   in external works (90.51)
Belt conveyors (B3j)
   excavating belt conveyor loaders (B2k)
Bench (law court facilities) 317
Benches (72.6)
   in external works (fixed and loose) (90.7)
   loose furniture and equipment (82)
Benches (construction plant) (B5j)
Bending forces (J4)
Bending machines (B5g)
Bending operations (D4)
Benefits, economic (Y6)
Berths 138
Beryllium bronze h6
Bibliography (Agm)
Bidets (74.24)
Biennials (plants) W
Billiards facilities 562
Bills of quantities (A4s)
Bin pounds (waste disposal facilities) 973
Binders
   clay, gypsum, magnesia and plastics r
   lime and cement q1/q3
   paints v3
Bins (bulk goods storage facilities) 1841
   agricultural 2681
Bins (elements of construction projects) (99.24)
   chute and hopper refuse disposal (52.13)
   manual refuse disposal (52.14)
   refuse disposal in general (52.19)
Biological factors (L6)
Birds
   factors affecting works (L62)
   nuisance protection services (68.6)
   rearing and living facilities 4647
   requirements (U51)
Bitumen bonded asbestos cement f6
Bitumen polythene n2
Bituminised paper j2
Bituminous felt n2
BITUMINOUS MATERIALS s
Blankets (protective plant) (B1d)
Blast furnace portland cement q2
Blast furnace slag
   air-cooled p2
   foamed p3
Blasting equipment (B5h)
   rock (B2h)
Blind boxes (76.7)
Blind people
   see Physically handicapped people
Blind people's gardens 587
Blinds (76.7)
Blockboard i4
Blockhouses 3756
Blocks (building divisions) 992
BLOCKS AND BLOCKWORK F
   large blocks and blockwork G
Blocks and tackle (B3x)
Blood transfusion facilities 428

Board
   paper j2
   vegetable particles other than wood j3
   wood fibre j1
   wood particles j7
   see also Boards
Board rooms 917
Boarding school facilities 7178
Boards
   rigid sheets R
   sections H
   see also Board
Boat control facilities 136
Boat manufacturing facilities 2755
Boat repair facilities 137
   manufacturing facilities 2755
Boat storage facilities 137
Boathouses 137
Boating facilities 546
Boats
   house boats 87
   manufacturing facilities 2755
   plant (B2w)
   water transport facilitites 138
Body building facilities 562
Body building units (road vehicles) 127
Body factors (U41)
Bogeys (transport plant) (B3i)
Bogs 092
Boiler houses 971
Boilers (99.123)
   hot water central heating (56.4)
   hot water supply and central heating (56.4)
   hot water supply services (53.3)
   piped and ducted services in general (59)
Boilers, steam (B8h)
Boiling pans (culinary fittings) (73.4)
Bollards (external works) (90.3)
Bolts (securing devices) t7
   or use (99.954)
Bolts and nuts t6
Bonding materials t3
Bonuses
   project (or office) administration (A1mr)
Book binders 346
Booking halls 348
Book-keeping (A2)
Boom trenchers (B2n)
Booths 994
Bore (internal diameter) (F47)
Bored pile foundations (17.2)
Bored piling equipment (B2r)
Boring (construction operations) (D5)
Boring equipment (B5)
Borrowed lights (32.4)
Borrowing (finance) (A2u)
Borstals 3765
Botanical gardens 751
Bowl urinals (74.45)
Bowling alleys 563
Bowling greens 566
Box frame structures (2.1)
Boxing facilities 562
Bracings (99.78)
Brass h6
Breadth (F47)
Breakage (X7)
Break waters 1863
Breather type felts L
Breeze (aggregates) p3
Brick, crushed p2
Brick industries facilities 2771

BRICKS AND BRICKWORK F
Bridge type cranes (B3u)
Bridges 182
   rail 118
   road 128
Bridleways 1233
Briefs (A3r)
Brightness (N5)
Brilliance (N5)
British Standards Specifications (Ajr)
Broadcasting facilities 151
Bronze h6
Browning work (thick coatings) P
Browsing facilities 768
Bucket elevator loaders (B2k)
Bucket wheel excavators (B2j)
Budgets (A2r)
Builders
   see Construction companies
   Construction personnel
'Builders' ironmongery' t6
Builders' merchants 348
Builders' yards 278
Building board, fibre j1
Building Centres 758
Building construction (E4)
BUILDING FRAMES (28)
Building plus external works (--)
Building Regulations (A3j)
Building Research Establishment Digests (Ajt)
Building schools (Ano)
Building science (E)/(Y)
Building societies facilities 336
Building systems (--)
Buildings (generally) 987
Built-in fittings (as a whole) (7-)
Built-up areas 052/058
Bulbs (plants, seeds) W
Bulk cement handling and storage plant (B4k)
Bulk goods storage facilities 184
   agricultural 268
Bulk-buying stores 341
Bulky organic manures w6
Bull dozers (B2g)
Bunk rooms 921
Bunks (72.1)
   loose furniture and equipment (82)
Burglar detection and alarm services (68.2)
Burners (99.365)
   gases supply services (54.9)
Bürolandschaft 32
Bursting (J4)
Bus stations 124
Bus systems 12
Buses 128
   transport plant (B3f)
Bushes W
Business activities (T132)
Butane gas supply (54.1)
Butt jointed tile work S
Butt joints Z
Buttressed retaining walls (16.2)
Buttresses (99.73)
   external walls (21.9)
Butyl n5
BUYING (A6)

Consult the schedules to check the context of items in the index. Italic entries are explained in the introduction.

Buz–Cir

Buzzers (99.125)
    audio communications (64.38)
    electrical services in general (69)
    fire detection and alarm (68.52)
    intruder detection and alarm (68.2)
    security and control services in general (68.9)
Bye-laws (Ajn)
    designing and physical planning (A3j)
By-passes 1223

C

CPM (A5t)
Cable and column 'frames' (28.5)
Cable bridges 1824
    cable braced (stayed) bridges 1828
    cable lift bridges 1825
Cable structures (2.6)
Cable transport facilities 113
Cables (elements of construction works) (99.362)
    communications services (64.9)
    electrical services in general (69)
    electric supply services (61.9)
    power subcircuits (62)
    lighting subcircuits (63.9)
    security and control services (68.9)
    transport services (66.9)
CABLES AND CABLEWORK (FORMS, CONSTRUCTIONS) J
Cafes 515
Caissons (16.3)
Calcined limestones q1
Calculators (Ago)
Calorific value (K42)
Camps
    defence 3755
    holiday 588
    settlements 056
Canal transport facilities 133
Candle power (N7)
Canoeing facilities 546
Canopies (27.7)
Canteens 511
    site accommodation (B1b)
Cantilever bridges 1828
Cantilever retaining walls (16.2)
Cantilevered roofs (27.7)
Cantilevers (99.742)
    in galleries, balconies (23.9)
Capacitance (Q5)
Capacity
    power (R3)
    supply factors (Y8)
    thermal (M4)
    volume (F45)
Capillarity (L35)
Capital
    cost (Y3)
    financing (A2s)
Capitols 312
Car parks 125
Car ports 963
Caravan sites 06:87
Caravans, residential 87
Carbon based refractory materials g6
Carbon dioxide fire fighting services (68.53)
    automatically operated (68.54)
Carbon fibres m9
Carborundum aggregate for granolithic and terrazzo q5
CARCASS (PRIMARY ELEMENTS) (2–)
Cardboard j2

Caretakers' houses 841
Carpets and carpeting work T
Carrels 768
Carriage sheds, railway 117
Carriageways 128
Carrying (construction operations) (D3)
Cars 128
    personnel (project administration) (A1mr)
    transport plant (B3f)
Case law (Ajk)
Cased and uncased pile foundations (17.2)
Casement windows (31.42)
Cash and carry stores 341
Casing work R
Casinos 581
CAST IN SITU WORK E
Cast iron h1
Cast stone f3
Casting (construction operations) (D2)
    equipment (B2d)
CASTING WORK E
Castles 86
Castor oil v3
Casualty facilities 418
Cat rearing and living facilities 4642
Catalysts u2
Catches t7
    or use (99.952)
Catering facilities 93
Catering fittings (73)
CATHEDRALS 62
Cattle grids (90.41)
Cattle rearing and living facilities 4644
Causes of change (X8)
Caves 094
Cavity floors (33.1)
Cavity internal walls (22.1)
Cavity wall insulation (21.1)
Cavity wall ties (21.1)
Cavity walls (21.1)
Cedar i2
CEILING FINISHES (45)
Ceiling openings (35.3)
Ceiling roses (lighting services) (63.9)
Ceiling walkways (35.8)
Ceilings
    applied ceiling finishes (45)
    suspended ceilings (35)
Cellars 993
    wine storage 935
Cellular bricks F
Cellular concrete, lightweight q6
    precast f4
Cellular glass o9
Cellular plastics n7
Cellular structures (2.3)
Cellulose derivatives n6
Celsius temperature (M7)
Cement q2
    asbestos cement f6
    asbestos cement mixes q9
    lime–cement q3
    lime–cement–aggregate mixes q4
Cement handling and storage plant (B4k)
Cement industries facilities 2771
Cement paints v8
Cemented coatings u1
Cement–lime mortars q4
Cement–sand mix q2
Cement–synthetic resin emulsions r4
Cemeteries 67
Cenotaphs 984
Centering (B2d)
Centigrade temperature (M7)
Central air conditioning (57.1)

Central dictation services (audio communications) (64.34)
Central government organisations (Akn)
    design and construction departments (Aks)
Central heating services (56.3)/(56.7)
    hot water, steam (56.4)
    warm air (56.5)
    electrical (56.6)
Central mechanical ventilation (57.6)
Centralised services (5.7)
Centrifugal pumps (B2t)
Ceramic wool m1
Ceramics g2
Ceremonial suites 318
Certificates
    agrément (Ajs)
    contract administration (A4x)
Cesspools 1741
Chain link J
Chains J
Chairs (72)
    loose furniture and equipment (82)
Chancels 68
Change (X)
Changing rooms 947
    entertainment facilities 528
Channels (elements of construction projects) (99.361)
    drainage services (52.9)
Channels (forms)
    pipes I
    sections H
Chantries 63
CHAPELS 63
    side chapels 68
Chapter houses 63
Charters (Ajw)
Charts (production diagrams) (A5v)
Chateaux 86
Chemical closets (sanitary fittings) (74.44)
Chemical conversion coatings u1
Chemical factors (L5)
Chemical wastes disposal (52.4)
Chemicals manufacturing facilities 274
Chemotherapy facilities in hospitals 4177
Child clinics 422
Children (U31)
Children's hospitals 416
Chilled water supply (53.1)
Chilling facilities 971
Chimes (audio communication) (64.38)
Chimneys as part of carcass, including flues (28.8)
    flues alone (59)
    free standing chimneys 161
China, vitreous g3
Chipboard j7
Chippers (B5d)
Chips (granular solids) Y
Chlorinated rubber n5
Chlorosulphonated polyethylene n5
Choirs 68
Chroma (G5)
Chromium h9
Chromium plated coatings u1
Chronic invalids welfare facilities 443
Church halls 63
CHURCHES 63
Chute and hopper refuse disposal (52.13)
Chutes I
Ciment fondu q2
Cinemas 525
Circuit breakers (99.416)
    electrical supply services (61.9)

Circuits (99.362)
  communications services (64)
  electrical services as a whole (6–)
  electrical supply services (61)
  lighting subcircuits (63)
  power subcircuits (62)
  security and control services (68)
  transport services (66)
Circulating activities (T191)
Circulation facilities 91
CIRCULATION FITTINGS
  as a whole (71)
  built-in or otherwise fixed (71)
  loose furniture and equipment (81)
Circulation pipes I
Circuses 526
Cisterns (99.23)
  liquids supply services in general (53.9)
  sanitary disposal fittings (wc's, urinals etc) (74.4)
Cities 053
  new cities, garden cities 054
City halls 314
City regions 052
City walls 3756
Civic centres 314
Civil courts 317
Civil defence facilities 3754
CIVIL ENGINEERING FACILITIES 1
Civil engineers (Amw)
Civil law enforcement facilities 374
Cladding work
  rigid sheet overlap N
  rigid sheet R
  section H
Claddings (99.81)
  finishes to structure (49)
  roof finishes (47)
  wall finishes, external (41)
Clam shell (drag and clam shell) (B2j)
Clamp-on vibrators (B6g)
Clamps t6
Classification (Ahi)
Classrooms 7291
  infants schools 712
  nursery schools 711
  schools 718
CLAY g
Clay industries facilities 2771
Clay mortar mixes r1
Clay soils p1
Clay–bitumen mixes s5
Clean air (H17)
Cleaners, dry 346
Cleaning
  construction operations (D8)
  ease of cleaning (V8)
  equipment for cleaning (B8u)
  routine maintenance (W2)
  use of facilities (T195)
CLEANING AND MAINTENANCE FITTINGS
  as a whole (75)
  built-in or otherwise fixed (75)
  loose furniture and equipment (85)
Cleaning facilities 95
Cleaning materials w7
Cleanliness factors (U464)
Clear glass o1
Clearways (air transport) 148
CLEARANCE OF SITE (D2)
  plant (B2)
CLEARING UP (D7)
Cleats t6

Clients (Amt)
  instructions (A3r)
  roles: project administration (A1mt)
Cliffs 092
Climate (H121)
Climbers (plants) W
Climbing tower cranes (B3t)
Clinics 422
  animal clinics 462
  special clinics 424
Clinker p3
Clips t6
Cloakroom fittings (76.6)
Cloakrooms 961
Clock systems (64.5)
Cloisters 68
Close contact adhesives t3
Closed circuit television facilities 153
Closed prisons 3762
Closets (sanitary fittings) (74.44)
Closing devices t7
  or use (99.961)
Clothing, protective (B1c)
Clothing manufacturing facilities 276
Clubs 534
  commercial 537
  health 421
  residential 536
Cluster form planning 084
Coach stations 124
Coagulants u2
Coal merchants 348
Coal mines 166
Coal tar s1
Coal tar pitch s1
Coarse aggregates
  see Aggregates
Coarse stuff q1
Coastal areas 092
Coastal protection facilities 1863
Coastal transport facilities 131
Coastguard stations
  protective service facilities 371
  water transport facilities 136
Coated macadam s5
Coated macadam manufacturing plant (B4f)
Coated macadam spreaders (B6c)
Coating (D6)
  ease of coating (V6)
COATING WORK
  FILM COATINGS V
  THICK COATINGS P
Coatings, protective u1
Cob g1
Cobblers 346
Cobbles and cobble paving work F
Codes of conduct (Ajw)
Codes of practice (Ajr)
Coding (Ahi)
Coffee bars 515
Cofferdams 187
  temporary works (B2r)
Cohesiveness (V6)
Coir j3
COLD (PROCESSES AND PROPERTIES) (M)
Cold storage facilities 965
Cold storage fittings (76.6)
  culinary display fittings (73.5)
Cold supply facilities 971
Cold water supply services (53.1)
Cold weather (H121)
Coldrooms 965
Collapse (J7)

Collapsible gates
  in external wall openings (31.54)
  in external works (90.3)
COLLEGE FACILITIES 72
  sixth form colleges 714
Colleges of Further Education 722
Collieries 166
Colour (G5)
Coloured cement q2
Coloured opaque glass o3
Coloured plain glass o1
Coloured translucent glass o2
Column and beam frames (28.2)
Column and beam structures (2.2)
Column and cable 'frames' (28.5)
Column and slab frames (28.3)
Column, beam and slab frames (28.2)
Columns (99.751)
  in building frame (28.9)
Combed joints Z
Combustibility (K41)
Combustible gas supply services (54.1)
Combustion products (K5)
Comfort (U45)
Commercial activities (T133)/(T134)
Commercial clubs 537
COMMERCIAL FACILITIES 33
COMMERCIAL FACTORS (Y)
Commercial lending libraries 762
Commercial organizations (Ako)
  design and construction (Aky)
Commercial use (H7)
Commissioning (A7q)
Commissions
  project (or office) administration (A1fp)
Committee rooms 917
Committees (Agj)
  project (or office) administration (A1gj)
COMMON AREAS 09
Common bricks F
COMMON FACILITIES 9
Common rooms 922
Commons 0871
Communal heating services (56.2)
Communal residential facilities 85
Communes 056
Communication cybernetics (Af)
  project (or office) administration (A1f)
COMMUNICATIONS (Ag)
  project (or office) administration (A1g)
COMMUNICATIONS FACILITIES 15
  common facilities 975
COMMUNICATIONS SERVICES (64)
  project (or office) administration (A1s)
Communities (settlement areas) 05
Communities (user groups) (U1)
Community centres 532
Community school facilities 713
Community services (piped, ducted, electrical) (5.7)
Compacting (D6)
Compaction equipment (B6e)
Compactors (refuse disposal) (52.18)
Companies, construction (Aky)
Compartment floors (23.8)

Consult the schedules to check the context of items in the index. Italic entries are explained in the introduction.

Com–Cou

Compartment walls
    external (21.8)
    internal (22.8)
Compartments (building divisions) 992
Compatibility (R7)
Competition and competitions (Afp)
    project (or office)
    administration (A1fp)
Completing (D7)
Completion (project
    administration) (A8r)
COMPLETION OF STRUCTURE (3–)
Compo mortars q4
COMPONENTS AND COMPONENTS
    WORK X
Composite door/window openings
    in external walls (31.3)
    in external walls (32.3)
    in roofs (37.3)
Composite floors (23.4)
COMPOSITE MATERIALS y
Composite pipes I
Composite storage fittings (76.1)
COMPOSITION (E)
    appearance (G1)
Comprehensive development areas
    see Action areas
Comprehensive school facilities 713
Compressed air supply (54.3)
Compression (J4)
Compressors (99.131)
    construction plant (B8f)
Computer facilities 766
Computers (Ago)
    cost planning (A4go)
    production planning (A5go)
Concave joints Z
Concentration (X1)
Concentric growth areas 084
Concert halls 522
Concourses 9141
Concrete q4
    crushed p2
    lightweight q6/q7
    precast f1/f5
Concrete plant
    compaction (B6g)
    manufacture (B4g)
    paving (B6d)
    transport and haulage (B3)
Condensation (L27)
Condensation resisting materials u3
Condensers (99.132)
Conditions of purchase (Y3)
Conditions of service (A1mq)
Conduct (Afn)
    codes (Ajw)
    project (or office)
    administration (A1fn)
Conductivity
    electrical (Q3)
    thermal (M33)
Conductors (99.362)
    electrical services (69)
Conduits (elements of construction
    projects) (99.361)
    communication services (64.9)
    electrical services in general (69)
    electrical supply services (61.9)
    lighting subcircuits (63.9)
    power subcircuits (62)
    security and control services (68.9)
Conduits (forms) I
Conference halls 916
Conference organizations (Akq)
Conferences (Anp)
Confessing facilities 68

Congress halls 916
Connection factors
    ability to accept fixings (E7)
    ease of connecting (V6)
Connector joints Z
Connectors t6
Conoidal roofs (27.5)
Consents (designing and physical
    planning) (A3j)
Conservation (control of change) (X8)
    economy in resources (T6)
    energy (R5)
Conservation planning areas 081
Conservatories 996
Consistency (V6)
Consortia (Aks)
Constituents (E1)
Constitution
    project (or office)
    administration (A1fg)
Construction (assembly) (E4)
CONSTRUCTION ADMINISTRATION
    AND MANAGEMENT (A)
Construction companies (Aky)
Construction industry facilities 278
CONSTRUCTION PLANT (B)
Construction technology (E)/(Y)
CONSTRUCTIONS AND PRODUCT
    FORMS A/Z
Consulates 316
Consultants
    project (or office)
    administration (A1my)
Consulting facilities
    common facilities 928
    medical facilities 423
Consumption (social economics) (U2)
Consumption of resources (T6)
    fuel and energy (R3)
Contact adhesives t3
Containers (99.2)
    site protection (B1d)
Contamination factors (L7)
Content (composition) (E1)
Contingencies (A9u)
Continuous access floors (33.1)
Continuous loaders (B2k)
Contract services (Y9)
Contraction (thermal) (M5)
Contraction joints Z
Contracts (A4u)
    documents (A4h)
    law (Ajk)
Contracts of service
    project (or office)
    administration (A1mq)
CONTROL AND SECURITY
    SERVICES (68)
Control devices (99.4)
    space heating services (56.9)
    air conditioning and ventilating
    services (57.9)
    transport services (66.9)
    control services (68.7)
Control gear (99.416)
    electrical supply services (61.9)
Control rooms 975
Control services (68.7)
Control towers, air traffic 146
Controlled environment agricultural
    facilities 2685

Controls (Aj)
    accounting (A2j)
    buying (A6f)
    cost planning (A4j)
    designing and physical planning (A3j)
    financial (A4w)
    importing (A6j)
    progress (A5f)
    project (or office) administration (A1j)
    quality (A7j)
    site and manufacturing (A5j)
    stock (A6s)
Conurbations 052
Convalescent homes 442
Convection (heat) (M34)
Convector heaters (56.9)
Convenience (U45)
Conveniences 941
Convention (U2)
CONVENTS 66
Conversion (W6)
Conveyors (construction plant) (B3j)
Conveyors (transport services) (66.5)
Cookers (73.4)
Cooking facilities 934
Cooking fittings (73.4)
    loose furniture and equipment (83)
Coolers (air treatment plant) (57.5)
Cooling towers 161
Co-operation (Af)
    project (or office) administration (A1f)
Co-ordination (administrative) (Af)
    project (or office)
    administration (A1fx)
Co-ordination (dimensional) (F43)
Copal v4
Copings (21.9)
Copper h5
Copper alloys h6
Copying of documents
    project (or office) administration (A1t)
Copyright (Ajx)
Cords J
Core (building division) 994
Cored structures (2.8)
Cork j5
Corms (plants, seeds) W
Corner joints Z
Corner shops 345
Cornices (ceiling finishes) (45)
Corridors 9141
Corrosion (L44)
    anti-corrosive materials u1
Corrosive wastes disposal (52.4)
Corrugated paper j2
Corrugated sheets and tiles N
Cosmic radiation (Q7)
Cost analysis (A4v)
Cost limits and yardsticks (A4j)
COST PLANNING AND CONTROL (A4)
    integrated financial/costs
    systems (A2)+(A4)
Costing (A4t)
Costs (Y2)
    costs in use (Y4)
    project (or office) administration (A2t)
    see also adjoining entries eg
    Cost planning and control
Cottage hospitals 412
Cotton j3
Couches (72.2)
    loose furniture and equipment (82)
Coumarone resins n6
Counterbalanced bridges 1825
Counterfort retaining walls (16.2)
Counters (catering fittings) (73.7)

Country planning areas 051
County courts 317
County offices 314
County planning areas 033
Couplings t6
Coursed blockwork F
Courses (education) (Anp)
Courses, sports 568
Court rooms 317
Courts, law 317
Courts, sports 568
Courtyards 998
Covered spaces 996
Covered swimming pools 541
Coverings (99.8)
   finishes to structure *(49)*
Coves (ceiling finishes) (45)
Cracking (J4)
Cranes (construction plant) (B3)
Cranes (transport services) (66.7)
Crash barriers (99.537)
   for motorways 121
Crawler mounted cranes (B3q)
Crawler mounted tower cranes (B3t)
Crawler tractor shovels (B2m)
Crawler tractors (B3g)
   earth-moving plant (B2f)
Creches 446
Crematoria 67
Creosote u3
Crib retaining walls (16.2)
Cricket facilities 564
Criminal courts 317
Criminal law enforcement facilities 374
Crippled people
   see Physically handicapped people
Critical path method (A5t)
Cross wall structures (2.1)
Cross walls (22.1)
Crossings (roads) 128
   level crossings 149
Cross-linking agents u2
Crowds (U1)
Crushed brick p2
Crushed concrete p2
Crushed stone p1
Crushers (refuse disposal) (52.18)
   construction plant (B4d)
Cryogenic temperature (M7)
Crypts 68
Cubic content (F45)
Cubicles 994
Culinary activities (T193)
Culinary facilities
   common facilities 93
   social refreshment facilities 51
Culinary fittings
   as a whole *(73)*
   built-in or otherwise fixed *(73)*
   loose furniture and equipment *(83)*
Culs-de-sac 1224
Cultivation work W
Cultural centres 532
Culverts (civil engineering) 181
Culverts (elements) (11.5)
Cupboards (76.2)
   airing cupboards (75.3)
   culinary storage fittings (73.5)
Cupolas (27.5)
Curing (properties and processes) (X6)
Curing agents u2
Curling facilities 567
Current electricity (Q11)
Currents, air (H122)
Curtain tracks (76.7)
Curtain walls (21.4)
Curtains (76.7)

Curved stairs (24.4)
Customs (U2)
Customs houses 315
CUTTING (D4)
   machines (B5f)
   machines (heat processes) (B5e)
   working properties (V4)
Cuttings (ground relief) (11.1)
Cybernetics (A8)
   project (or office) administration (A1f)
Cycle racing facilities 564
Cycle sheds 963
Cycle tracks 1232
Cyclones (H122)
Cylinders H
Cylindrical roofs (27.5)

D

Dairies 273
Damage (X7)
   accidental damage (L61)
   fire and explosions (K7)
Damp proof courses (99.532)
   in floor beds *(13.9)*
   in external walls *(21.9)*
Damp proof sheets and strips
   flexible, proofing L
   malleable M
Damp proofing (L34)
Dampness (L31)
Dams 187
Dance halls 521
Dangerous buildings 988
DARK (N)
Dark rooms 732
Data communications facilities 1542
Data processing (Agn)
   cost planning (A4gn)
   production planning (A5gn)
Data processing facilities 766
Data transmission services
   (communications) (64.44)
Day patients facilities 418
Daylight (N11)
   factor (N7)
Dayworks (A4w)
Dead loads (J4)
Dead people disposal facilities 177
Dead weight rollers (B6f)
Deaf people
   see Physically handicapped people
Deal (sections) H
Dealers (bulk materials) 348
Deaneries 61
Debating chambers 318
Debts (finance) (A2t)
Decay (quality change) (X7)
   biological factors affecting works (L6)
Decibels (P7)
Decision making (Afi)
   project (or office)
     administration (A1fi)
Decking (99.52)
   for floors *(23.9)*
   for roofs *(27.9)*
Decking work R
Decline (X6)
Declining areas 0833
Decorations (48)
Decorative buildings 984
Decorative coating work V
Decorative sheets and tiles T
Deep freeze storage 965
Defective buildings 988
Defective work (A7r)

Defects in use (T8)
Defects procedure (A8r)
Defence facilities 375
Deficiency in use (T8)
Deformation (J7)
   resistance (J3)
Degreasing facilities 127
De-ionised water supply (53.6)
Delays (production planning) (A5r)
Delicate people
   see Handicapped people
Deliquescence (L23)
Delivery
   stock for construction projects (A6u)
   supply factors (Y8)
Deltas 093
Demography (U2)
Demolishing (D2)
Demolition plant (B2b)
DEMOLITION WORK B
   used in specifications applications
Demountability (V2)
Demountable buildings 983
Demountable partitions (22.3)
Density (high density areas) 085
Density (physical properties) (F5)
Dental hospitals 414
Dentist's surgeries 424
Department stores 343
Departments 992
Departments, government (facilities)
   local 315
   national 312
Departments, government
   (organisations) (Akn)
Deposition (L44)
Depreciation (A2t)
Depressed areas 0833
Depressions (ground relief) (11.1)
Depth (F47)
Derelict areas 0834
Dereliction (X7)
Derricks (B3r)
Design, graphic (Ahm)
Design and construction
   organizations (Akr)
Design centres 758
Design colleges 722
Design offices 32
   common facilities 926
Design personnel (Amw)
   project (or office)
     administration (A1mw)
Design process (A3f)
Design team meetings (A3gj)
Designated areas 081
Desks (72.3)
   loose furniture and equipment (82)
Destitute people (U36)
   welfare facilities 445
Detachability (V2)
Detached buildings 982
Detached housing
   single storey 8111
   two storey 8121
Detail design (A3)
Detection and alarm services (68.1)
   against fire (68.52)
   against intruders (68.2)

Consult the schedules to check the context of items in the index. Italic entries are explained in the introduction.

Det–Dur

Detectors (99.4)
    security and control services (68.9)
Detention facilities 376
Detergents w7
Deterioration (quality change) (X7)
    factors affecting works (L8)
Development (change) (X6)
DEVELOPMENT (RESEARCH) (Ao)
Development planning areas 081/083
Development plans
    see Structure plans
Development sites (– –)
Devolution (X6)
De-watering equipment (B2t)
Diagnostic activities (T141)/(T142)
Diagnostic facilities
    hospitals 4171
    medical buildings 42
Dialling codes, telephone (A1s)
Diameter (F47)
Diaphragm retaining walls (16.2)
Diatomite p1
Dictation services (audio
    communications) (64.34)
Dielectric constant (Q2)
Differentiation (X2)
Diffusability, gas (L24)
Diffuse reflecting glass u2
Diffuse reflection factor (light) (N41)
Diffuse transmission (light) (N3)
Digests, BRE (Aju)
Digging plant (B2e)
Dilapidated buildings 988
Dilapidation (X7)
Dimensional change (F47)
Dimensional co-ordination (F43)
Dimensional systems (F7)
Dimensions (F47)
Dining facilities
    common facilities 937
    social refreshment facilities 51
Dining tables (73.7)
    loose furniture and equipment (83)
Direct costs (A2t)
Direct reflection factor (light) (N41)
Direct space heating appliances
    (independent) (56.8)
Direct works organisations (Aks)
Directors
    project (or office)
    administration (A1my)
Disabled people
    see Physically handicapped people
Discarding (X8)
Discharge lighting sources (63.9)
Discotheques 521
Discount trading stores 341
Discussion facilities 928
Disease
    biological change factors (L63)
    health and hygiene factors (U466)
Dish washing fittings (73.2)
    loose furniture and equipment (83)
Disintegration (X2)
Dismantling, ease of (V2)
Dispensaries
    animals 462
    hospitals 4177
Dispersing agents u2
Displacement pile foundations (17.3)
    Small displacement (17.4)
Display activities (T175)
Display facilities 75
Display fittings (71.2)
    loose furniture and equipment (81)
Display lighting (63.2)
Disposability (V8)

Disposal bays (air transport) 148
Disposal facilities
    common facilities 973
    public health engineering 174/178
Disposal fittings, culinary (73.2)
Disposal fittings, sanitary (74.4)
DISPOSAL SERVICES (52)
Disseminating information (Ag)
Dissociative change (X2)
Dissolving (L37)
Distance (F47)
Distempers V8
Distribution (dissociative charge) (X2)
Distribution (stock) (A6t)
*Distribution boards (electrical supply
    services) (61.9)*
Distribution boxes (B8e)
District heating services (56.2)
District offices 314
District planning areas 035
District services (5.7)
Disturbance (X4)
Dividends (A25)
Dividing elements (2.9)
    horizontal (2.9)
    vertical (21)
Diving pools 544
Divisions within buildings
    functional 992
    horizontal 993
    vertical 994
Docks (law courts) 317
Docks (water transport) 138
Docks, scenery 524
Doctors' surgeries 423
Documentation (Agm)
    cost planning (A4gm)
    production planning (A5gm)
    project (or office)
    administration (A1gm)
Documentation preparation (Ah)
    cost planning (A4h)
    designing and physical
    planning (A3h)
    production planning (A5h)
    project (or office)
    administration (A1h)
Dog leg stairs (24.3)
Dog racing facilities 564
Dog rearing and living facilities 4642
Dome lights (37.41)
Domes (27.5)
Domestic animals (requirements) (U51)
Door and window furniture (31.9)
Door frames (31.59)
Door furniture (31.59)
Door sets (31.5)
    in internal openings (32.5)
Door sills (31.59)
Doors (31.5)
    internal doors (32.5)
Doorways (31.5)
    internal doorways (32.5)
Dormer windows (37.44)
Dormitories 921
Dormitory towns 053
Double-lap tiles and tiling work N
Dovetail joints Z
Dowelled joints Z
Downland 092
Dozers (B2g)
Drag and clam shell (B2j)
Drag line (B2j)
DRAINAGE AND WASTE DISPOSAL
    SERVICES (52)
    drainage in external works (90.51)
    land drainage (11.5)
Drainage facilities 187

Draining boards (culinary fittings) (73.2)
Drama facilities 524
Drapes (76.7)
Draughts (internal environment) (H6)
Drawers (storage fittings) (76.3)
    loose furniture and equipment (86)
Drawing (A3t)
Drawing offices 32
Drawings (A3t)
Drawings facilities 765
Dray shovels (B2j)
Dredgers (B2w)
Drenchers (fire fighting services) (68.54)
Dressing (construction operations) (D4)
Dressing rooms
    common facilities 947
    entertainment facilities 528
Dried clay g1
Driers (99.16)
    air treatment plant (57.5)
    *cleaning and maintenance
    fittings (75.3)*
    hand drying fittings (74.3)
Drilling (D5)
    ease of drilling properties (V5)
    equipment for rock drilling (B5c)
    machines (B5f)
Drink manufacturing facilities 273
Drinking fountains (73.8)
Drinking water supply (53.1)
Drive ins 128
Driven pile foundations (17.3)
Drives, road 1224
Driving (construction operations) (D6)
    pile driving equipment (B2r)
Droughts (H16)
Dry carriage refuse disposal (52.13)
Dry chemical fire fighting services (68.53)
    automatically operated (68.54)
Dry cleaners 346
Dry cleaning fittings (75.8)
Dry docks 138
Dry risers (fire fighting service) (68.55)
Dry rot (L64)
Dry weather (H121)
Drying agents u2
Drying equipment (B8i)
Drying facilities 953
Drying factors (L36)
    working factors (V6)
Drying fittings
    cleaning and maintenance
    fittings (75.3)
    sanitary and hygiene fittings (74.3)
Drying oils v3
Dryness (L31)
Dual carriageway roads 1225
Ducted services (5–)
*Ducts (elements of construction
    projects) (99.361)*
    *air conditioning and ventilation
    services (57.9)*
    as part of carcass (28.8)
    as secondary elements (38)
Ducts and duct work (forms,
    constructions) I
Dump trucks (B3e)
Dumpers (B3e)
Dumps, bulk goods storage 1848
Dumps, refuse 1753
Dunes 092
Durability (R8)
Durable goods shops 345
Duralumin h4

Dus–Exp

Dust (pollutant) (L7)
Dustbins (52.14)
Dutch barns 2684
Dwellings 81
    for special classes of user 84
    historical 86
    individually designed 82
    temporary or mobile 87
Dyes v2
Dykes 187
Dynamics factors (J)

E

EEC Commission facilities 311
ESN
    see Handicapped people
Ear, nose and throat hospitals 414
Earnings (A2s)
Earth p1
Earth closets (sanitary fittings) (74.44)
Earth compaction equipment (B6f)
Earth drilling equipment (B5c)
Earthenware g2
Earth-moving plant (B2f)
Earthquake stations 737
Earthquakes (H16)
Easy chairs (72.2)
    loose furniture and equipment (82)
Eating activities (T193)
Eating facilities
    common facilities 937
    social refreshment facilities 51
Eaves (roof secondary elements) (37.6)
Ebonite n5
Echo (P4)
Ecological factors (H17)
Economic factors (Y1)
    characteristic of areas 096
    social (U2)
Economic resources (A2)
Economic status of people (U36)
Ecumenical centres 61
Edge joints Z
Edgings (99.98)
    finishes to structure in general (49)
    floor finishes (43)
    ground surface treatments (external works) (90.41)
    roof finishes (47)
    stair finishes (44)
EDUCATION AND TRAINING (An)
EDUCATIONAL FACILITIES 71/73
Educational use (H7)
Educationally subnormal people
    see Handicapped people
Eelgrass j3
Effectiveness (T3)
Effects of change (X8)
Efficiency
    economic (Y7)
    energy conversion (R3)
    in use (T3)
Efflorescence (L44)
Effluent treatment and disposal (52.4)
Effluents (pollutants) (L7)
Egress facilities 913
Elasticity (J7)
Elastomers n5
Elderly people
    see Old people
Electric blankets (plan) (B1d)
Electric lighting (63)

Electric tractors, battery operated (B3g)
    earth-moving plant (B2f)
Electrical central heating (56.6)
Electrical engineering manufacturing facilities 2754
ELECTRICAL FACTORS (Q)
Electrical services (6-)
ELECTRICAL SUPPLY SERVICES (61)
Electricity (Q1)
    as power source for services (5.14)
Electricity generating stations 162
Electricity meters (61.9)
Electricity supply facilities 162
    common facilities 971
Electro acoustics (P1)
Electromagnetic radiation (Q7)
Electromagnetism (Q6)
Electromotive force (Q5)
Electroplated coatings u1
Electrostatically charged coatings u1
ELEMENTS (of construction projects) (1−)/(9−)
    ground and substructure elements (1−)
    structure primary elements (2−)
    structure secondary elements (3−)
    structure finishes elements (4−)
    services elements (5−)/(6−)
    fittings elements (7−)/(8−)
    external works elements (90)
Elevator loaders, bucket (B2k)
Elevation (F47)
Elm i3
Embankments (11.1)
Embarkation facilities
    air transport 144
    common facilities 149
    rail transport 114
    road transport 124
    water transport 134
Embassies 316
Embedded membrane work L
Emergency electrical supply (61.7)
Emergency lighting (63.8)
Emergency posts, first aid 426
Emergency signs (71.1)
Emission (X2)
    light (N3)
    thermal (M37)
Employees (U36)
Employers (U36)
Employment (A1mn)/(A1mq)
Employment exchanges 315
Emulsifying agents u2
Emulsion paints v6
Emulsions Y
    bitumen emulsion s1
    cement–synthetic resin emulsions r4
Enamels v4
Encampments 056
Enclosures (in external works) (90.3)
ENERGY (R1)/(R5)
Energy conservation (R5)
Energy consumption (R3)
Energy demand (R3)
Energy resources (U6)
Engagements of personnel (A1mq)
Engineering bricks F
Engineering industries facilities 275
ENGINEERING WORKS 1
Engineers (Amw)
    central and local government departments (Aks)
    works by particular engineers 987
Enquiries, public (Ajy)
Enquiry facilities 768

Entertainment activities (T151)/(T153)
Entertainment arenas 521
ENTERTAINMENT FACILITIES 52
Entrance halls 913
Entrance mats (71.3)
Entrance screens (31.3)
Entrances (doorways) (31.5)
ENVIRONMENT (H)
    human response (U48)
Environmental features 998
Episcopal palaces 61
Epoxy resins n6
Equestrian facilities 565
Equilibrium (X5)
Equipment, construction (B)
Equipment, office
    drawing (A3t)
    filing (A1u)
    library (A1gm)
    photocopying etc. (A1t)
    secretarial (A1s)
Equipment, site investigation and surveying (A3s)
Ergonomics (U41)
Erosion (L44)
Erosion protection facilities 1863
Escalator enclosures 9142
Escalators (66.4)
Escape
    fire safety (K24)
Escape stairs (24.5)
Estimating (A4t)
Estuaries 093
Ethics (U2)
    administration generally (Afn)
    project (or office) administration (A1fn)
Ethylene propylene n5
European parliament facilities 311
EVALUATING (Aq)
    price (A4t)
Evaporators (99.132)
Evenness (G4)
Evolution (X6)
Examination facilities, medical 428
Excavated and filled ground (11)
Excavating (D2)
    plant (B2j)
Excavating belt conveyor loaders (B2k)
EXCAVATING WORK C
Excavations (ground elements) (11.9)
Exchanges (communications services elements) (64.32)
Exchanges (communications facilities)
    switching (telecommunications) 156
    telephone 1541
Exchanges, employment 315
Exchanges, stock 337
Executive offices 32
Exfoliated vermiculite p3
EXHIBITION FACILITIES 75
Exhibition organizations (Akq)
Exit vestibules 913
Exits (doorways) (31.5)
Expanded clays p3
Expanded mesh J
Expanded mica p3
Expanded perlite p3
Expanded plastics n7

Consult the schedules to check the context of items in the index. Italic entries are explained in the introduction.

Exp–Fle

Expanding towns 055
Expansion (dimensional change) (F47)
   thermal (M5)
Expansion joints Z
Expenditure (A2t)
Explosions (K6)
   damage (K7)
Explosives w8
Exposure (H123)
Extenders u2
EXTERNAL ELEMENTS (9–)
External envelope (2.9)
External spaces 998
EXTERNAL WALLS (21)
   secondary elements (31)
   external finishes (41)
EXTERNAL WORKS (90)
   Building plus external works (– –)
Extraction (X2)
Extractors (B2r)
Eyebrow windows (37.44)

F

Fabrics T
Facades of buildings 995
Face shovels (B2j)
FACILITIES 1/9
Facing bricks F
Facsimile transmission facilities 1542
Facsimile transmission services (64.43)
Factories 282
Fahrenheit temperature (M7)
Faience g3
Failure
   in use (T8)
   mechanics (J7)
Fairfaced blockwork and brickwork F
Fairgrounds 583
Fall-out shelters 3757
Families (U1)
Fans (99.33)
   air conditioning and ventilation services (57.9)
   construction plant (B8k)
Farm animals (requirements) (U51)
Farming facilities 263
Farms, sewage 1742
Fascias (99.87)
   roof finishes (47)
Fasteners t6
   or use (99.94)
Fathers (U32)
Fauna 0874
Feasibility studies (A3)
Feedback from projects (A8s)
Feeders (B4d)
Feel (G8)
Fees (A2s)
Felts T
   impregnated n2
   waterproofing L
Females (U33)
Fences (external works) (90.3)
Fences (site protection) (B1a)
Fencing facilities 562
Fenland 092
Fens (land reclamation) 185
Feretory spaces 67
Ferns W
Ferrous metals h1/h3
Fertilisers w6
Fibre building board j1

Fibres
   aggregates p7
   impregnated n2
   inorganic m
   mixed natural/synthetic n9
   synthetic n6
   vegetable and animal j
   vegetative products w6
Field events facilities 564
Field hospital facilities 418
Field paths 1234
Field posts, first aid 426
Fields 2688
Fighting sports facilities 562
Filament lighting sources (63.9)
Filing (A1u)
Fill (ground elements) (11.9)
Filled and excavated ground (11)
Fillers
   gap-filling adhesives t3
   joint fillers t4
   paint preparation fillers v1
Fillets H
Filling stations 126
Filling work C
Fills p
FILM COATING WORK V
Film projection facilities 768
Film projection services (64.12)
Filters (99.142)
   air treatment plant (57.5)
   drainage services in general (52.9)
   liquids supply services in general (53.9)
   piped and ducted services in general (59)
Final accounts (A4x)
Finance
   activities in facilities (T133)
   availability (Y1)
   integrated financial/costs systems (A2) + (A4)
   project (or office) administration (A2)
Financial control (A4w)
Fine aggregates
   see Aggregates
Fining coating work P
Finishes (description) (E1)
Finishes, external e.g. [(41) + (47)] (4–)
Finishes, internal e.g. [(42) + (43) + (45)] (4–)
FINISHES TO STRUCTURE (4–)
Finishing work
   flexible sheets and tiles T
   thick coating P
   top coating V
Fir i2
FIRE (K)
Fire barriers (99.533)
   in electrical services (69)
   in piped and ducted services (59)
   in structure (29)
Fire blankets (68.56)
Fire bricks F
Fire buckets (68.56)
Fire detection and alarm services (68.52)
Fire divisions 992
Fire extinguishers (68.56)
Fire fighting services (68.53)
   automatic operation (68.54)
   manual operation (68.55)
   portable systems (68.56)
Fire hose drying spaces 372
Fire practice towers 372
Fire protection (K2)
Fire protection and security services (68.1)

Fire protection services (68.5)
Fire resistance (structures) (K3)
Fire stations 372
Fireclay g6
Fireplaces (56.8)
Fired clay g2
First aid posts 426
First schools 712
Fish (requirements) (U51)
Fish breeding facilities 4641
Fish plate joints Z
Fisheries 262
Fishing facilities
   agricultural facilities 262
   sports facilities 565
Fit (F6)
FITTING (D4)
Fitting facilities (shops) 348
FITTINGS (fixed) (7–)
   in external works (90.7)
Fittings (loose) (8–)
   in external works (90.7)
Fittings (project or office administration) (A1r)
FITTINGS AS A WHOLE (fixed and loose) (7–)/(8–)
   code (7–) alone is usually used, particularly in project applications
Fittings for special activities (as a whole) (77)
Five storey housing 815
Fives facilities 562
Fixed fittings (as a whole) (7–)
Fixed joints Z
Fixed partitions (22.3)
Fixing
   construction operations (D6)
   construction plant (B6)
   ease of fixing (V6)
   factors relating to composition, content (E7)
Fixing devices t6
   or use (99.94)
FIXING MATERIALS t
Fixtures
   see Fittings
Flagpoles (71.1)
   in external works (90.7)
Flags (paving slabs) S
Flame retardants u4
Flame spread (K45)
Flammability (K45)
Flanged joints Z
Flash point (K44)
Flashings (99.98)
   finishes to structure in general (49)
   roof finishes (47)
   wall finishes (41)
Flashings work with malleable sheets M
Flat ground (11.1)
Flat roofs (27.1)
Flatness (G4)
Flats 816
Flatted factories 282
Flattening (D4)
Flatting agents u2
Flax j3
Flax board j3
Flettons F
Flexibility (adaptability) (T5)
Flexibility (mechanics) (J7)
FLEXIBLE PROOFING SHEETS AND SHEET WORK L

FLEXIBLE SHEETS AND SHEET WORK T
Flint aggregate granolithic q5
Floating buildings 983
Floating docks 138
Floating floors (43)
Floating work (thick coatings) P
Flocculants u2
Flood protection facilities 187
Flood protection services (68.6)
Floodlighting (63.6)
    in external works (90.63)
Floods (H16)
FLOOR BEDS (13)
    secondary elements (33.2)
    finishes (43)
FLOOR FINISHES (43)
Floor openings (33.5)
Floor polishing fittings (75.8)
    loose furniture and equipment (85)
Floors (spaces) 993
FLOORS AND GALLERIES (23)
    secondary elements (33)
    finishes (43)
Flora 0874
Flow factors (R1)
Flowering shrubs W
Fluctuations (contract
    administration) (A4x)
Flues (99.361)
    as part of carcass (28.8)
    piped and ducted services (59)
Fluid mechanics (J21)
Fluids (formless products) Y
Fluids (interactive factors) (L1)
Fluids (state of substances) z1
    source of energy (R1)
Fluorescent lighting sources (63.9)
Flush blockwork and brickwork F
Flush joints Z
Flush lighting fittings (63.9)
Flush pipes I
Flux, light (N7)
Fly ash p4
Fly galleries 524
Flyovers 128
Foam fire fighting services (68.53)
    automatically operated (68.54)
    manually operated (68.55)
    portable appliances (68.56)
Foam inlets (fire fighting services) (68.55)
Foamed blast furnace slag p3
Foamed plastics n7
Foaming agents u2
Foams Y
Fog (H122)
Foils L
Folded plate roofs (27.4)
Folded plate structures (2.4)
Folding partitions (32.54)
Folding shutter doors (31.54)
    in internal openings (32.54)
Food counters (catering fittings) (73.7)
Food manufacturing facilities 273
Food preparation and storage
    facilities 93
Food preparation and storage
    fittings (73)
Food shops 345
Foot baths (74.24)
Football facilities 564
Footbridges 1823
Footings (16.4)
Footpaths 1234
Footwear manufacturing facilities 2763
Forces (J4)
    electromotive force (Q5)

Forecasts
    financial (A2r)
    sales (A6f)
Forecourts 998
Forensic facilities 428
Forestry facilities 261
Forges 346
Fork-lift trucks (B3h)
Form (G1)
FORMED MATERIALS e/o
FORMING (D4)
FORMLESS MATERIALS p/s
FORMLESS WORK Y
Forms (formwork) (B2d)
FORMS (PRODUCTS) A/Z
FORMS OF PRESENTATION
    (DOCUMENTS) (Z)
Formwork (B2d)
Fortified houses 86
Fossil fuel extraction and supply
    facilities 164/166
Foul drainage (52.3)
Foundations (16.4)
FOUNDATIONS AND RETAINING
    WALLS (16)
Fountains 984
    external works (90.7)
    washing fittings (74.23)
Fountains, drinking (73.8)
Four storey housing 814
Foyers 913
Framed structures (2.2)
Frames (99.71)
FRAMES (BUILDING FRAMES) (28)
Frames, pile (B2r)
Framing and cladding internal
    walls (22.6)
Framing and cladding walls (21.6)
Freestanding chimneys 161
Freezers (culinary fittings) (73.5)
Friaries 66
Friction (J5)
Friendship (U2)
Frost (H122)
Frost resisting agents u2
Fruit trees W
Fryers (culinary fittings) (73.4)
Fuel consumption (R3)
Fuel extraction and supply
    facilities 163/166
Fuel gas supply services (54.1)
Fuels w3
Fume chambers 732
Fumes (K5)
Functional divisions 992
FUNCTIONAL MATERIALS t/w
Funeral vaults 67
FUNERARY FACILITIES 67
Funfairs 583
Fungi (L62)
Fungicides u3
    ancillary materials w9
Fur manufacturing facilities 2762
Furnishings, soft (78.3)
    curtains and drapes (76.7)
Furniture (fixed) (7−)
    in external works (90.7)
Furniture (loose) (8−)
    in external works (90.7)
Furniture, street 128
Furniture as a whole (fixed and
    loose) (7−)/(8−)
    code (7−) alone is usually used
Furniture manufacturing facilities 2772
Furrings H
Further Education facilities 72

Fuses (99.416)
    electrical services in general (69)
    electrical supply services (61.9)
    lighting subcircuits (63.9)
    power subcircuits (62)

G

G P hospitals 412
GRC f2
GRCS f1
GRG f7
GRP n8
Gable walls (21.8)
Gabled roofs (27.8)
Gales (H122)
Galleries (circulation facilities) 9141
Galleries (display facilities) 758
    art galleries 757
Galleries (elements of construction
    projects) (23.7)
Galleries (solid fuel mines) 166
Galleries, fly 524
Galleries, watching 68
GALLERIES AND FLOORS (23)
    secondary elements (33)
    finishes (43)
Galvanised coatings u1
Gambling facilities 581
Game (requirements for hunted animals,
    birds) (U51)
Gangways
    circulation facilities 9141
    water transport facilities 138
GANTT charts (A5v)
Gaols 376
Gap-filling adhesives t3
Garages
    common facilities 963
    road vehicle repair facilities 127
Garbage disposal services (52.1)
Garden cities 054
Gardens 998
    beer gardens 517
    botanical and zoological gardens 751
    gardens for the blind 587
    leisure gardens 587
    market gardens 264
    roof gardens 993
Garrets 993
Gas as power source (5.13)
Gas extraction and storage facilities 165
Gas fuels w3
Gas heating (56.13)
Gas lighting (63.8)
Gas meters (54.9)
Gas supply facilities 165
    common facilities 971
Gaseous waste disposal services (52.2)
Gases (formless products) Y
Gases (interactive factors) (L2)
    explosive limits (K6)
    mechanics (J22)
GASES SUPPLY SERVICES (54)
Gaskets t4

Consult the schedules to check the context of items in the index. Italic entries are explained in the introduction.

Gat–Hea

Gates (external works) (90.3)
    collapsible gates in wall
    openings (31.54)
Gauges (99.4)
    electrical services (69)
    piped and ducted services (59)
Gelling agents u2
Gels Y
General hospitals 412
Generating stations 162
Generators (99.126)
    construction plant (B8)
    electrical supply services (61.9)
Geodesic structures (2.4)
Geological factors (H16)
Geometry (F1)
Geophysical stations 737
Geotechnics (J12)
Geothermal heat production
    facilities 161
Geriatric clinics 422
Geriatric hospitals 416
German silver h6
Gin wheels (B3x)
Girders
    see Beams
Glare (N6)
    anti-glare glass o7
Glass o
Glass fibre reinforced calcium silicate f1
Glass fibre reinforced cement f2
Glass fibre reinforced gypsum f7
Glass fibre reinforced plastics n8
Glass industries facilities 2771
Glass wool m1
Glasshouses 264
Glazed fired clay ware g3
Glazed stoneware g3
Glazed terra cotta g3
Glazes v4
Glazing (99.86)
    for window openings (31.49)
Glazing units, multiple o5
Glazing work R
Gliding facilities 567
Glossaries (Agh)
Glues t3
Goats rearing and living facilities 4645
Gold h9
Golf courses 566
Goods lifts (66.1)
Government departments
    local offices 315
    national facilities 312
Government organizations (Akm)
Grabs (B3x)
Grade separated roads 1225
Graders (B2i)
Grading of personnel (A1mm)
Grammar school facilities 713
Grandstands 568
Granite e1
Granolithic q5
    precast f3
Granular aggregates, artificial p2/p3
Granules Y
Graphic design (Ahm)
Grass materials j3
Grasses W
Grassland 092
Gratings (99.58)
    drainage services (52.9)
Grave yards 67
Gravel p1
Gravity conveyors (66.5)

Gravity retaining walls (16.2)
Green belt 051
Green rooms 528
Green wedges 051
Greenhouses 264
Greens, bowling 566
Gridiron planning 084
Grids, cattle (external works) (90.41)
Grillages (16.4)
Grilles (99.63)
    air conditioning and ventilation
    services (57.9)
Grilles, roller (internal hatch
    openings) (32.75)
Grilles, tree (external works) (90.41)
Grills (culinary fittings) (73.4)
Grinders (refuse disposal) (52.18)
Grinding (D4)
    machines (B5f)
Gritstone e4
GROUND (11)
    composition (11.2)
    inclination (11.1)
    relief (11.1)
    underground (11.5)
GROUND AND SUBSTRUCTURE (1–)
Ground engineering (11)
Ground preparation (in external
    works) (90.1)
Ground surface treatments (in external
    works) (90.4)
    floor beds (13)
Ground underwater (13.6)
    in external works (90.43)
Ground water (11.4)
Group consciousness (U2)
Group heating services (56.2)
Group practices (Akv)
    medical facilities 423
Groups, user (U1)
Growth (X6)
    administration (Afh)
Growth inhibitors w9
Growth stimulants w6
Groynes 1863
Guarantees (Y3)
Guard rails (B1a)
Guesthouses 854
Guide rails (99.363)
    for lift shafts (34)
Guideways 113
Guiding devices (99.36)
Guildhalls 314
Gullies (99.365)
    drainage services (52.9)
Gun machines (B6j)
Gunmetal h6
Gunsmiths 346
Gusset plate joints Z
Gutters (forms) I
Gutters (rainwater drainage) (52.5)
Gymnasia 562
Gymnastic facilities 562
Gynaecological hospitals 415
Gypsum r2
    preformed f7
Gypsum plasters r2
Gypsy encampments 056

H

Hail (H122)
Hair j6
    seed hairs j3
Half brick brickwork F
'Half way housing' 87
Halls (assembly facilities) 916
Halls (rooms) 994
    entrance halls 913
Halls of residence 856
Halts, railway 114
Hamlets 051
Hammers (B2r)
Hand driers (74.3)
HAND TOOLS (B7)
Handbooks
    project (or office)
    administration (A1jw)
Handicapped people (U35)
Handicapped people's facilities
    common facilities 988
    housing facilities 848
    school facilities 7172/7173
    welfare facilities 444/445
HANDING OVER (A8)
Handling (D3)
Handrails (99.363)
    stair and ramp secondary
    elements (34)
Handsets (telephone services) (64.32)
Hangars 147
Harbours 134
Hard surfaces
    floor beds (13.1)
    in external works (90.41)
Hardboard j1
Hardcore p2
Hardeners u2
    surface hardeners u5
Hardness (J4)
Hardstandings (roads) 128
    external works (90.41)
Hardwood i3
Hatch openings (32.7)
    in external walls (31.7)
Hatches (99.63)
    in external walls (31.7)
    in internal walls (32.7)
Haulage plant (B3b)
Haulage systems 12
Hazards (U47)
Header bond blockwork and brickwork F
Heading joints Z
Health activities (T14)
Health clubs 421
Health factors (U46)
Health service facilities 41/42
Heart hospitals 414
Heart units (integrated services) (5.2)
Hearths (33.2)
HEAT (M)
    reaction to fire (K43)
Heat absorbing glass o6
Heat production facilities 161
Heat rejecting glass o6
Heat supply facilities 161
    common facilities 971
Heaters (99.123)
    air treatment plant (57.5)
Heaths 092
Heating and hot water supply (56.4)
Heating and ventilating services (56)/(57)
Heating appliances (independent) (56.8)
Heating mains (external works) (90.52)
Heating plant (B8i)

Hea–Inf

Heavy artificial granular aggregates p2
Heavy concrete q4
   precast f2
Heavy industry facilities 281
Hedge plants W
Height (F47)
Helical stairs (24.4)
Heliports 142
Helm roofs (27.23)
Helmets, protective (B1c)
Hemihydrate r2
Hemlock i2
Hemp j3
Herbaceous plants W
Herbaria 751
Herds (U1)
Hereditary groups (U1)
Hertz (sound measurement) (P7)
Hessian j3
Hiding power (V6)
High altitude areas 095
High alumina cement q2
High commissions 316
High density areas 085
High income people (U36)
High rise buildings 981
High school facilities 713
High temperature (M7)
High voltage electrical mains
   supply (61.6)
Highly-priced areas 096
Highway engineering 12
Hilly areas 092
Hinged joints Z
Hinges t6
Hipped roofs (27.23)
Hire services (Y9)
Historical buildings 986
   residential facilities 86
Hoardings (external works) (90.7)
Hoardings, road 128
Hobby facilities 928
Hockey facilities 564
Hoggin p1
Hoisting (D3)
   plant (B3p)
Hoists (construction plant) (B3w)
Hoists (transport services) (66.2)
Holding bays (air transport) 148
Holding power (J4)
Holiday camps 588
Holiday facilities 588
Holiday towns 053
Hollow blocks and bricks F
Hollow sections H
Hollows (ground relief) (11.1)
Homeless people (U36)
HOMES, WELFARE 44
Honeycomb bond blockwork and
   brickwork F
Hoods (culinary fittings) (73.4)
Hoppers (bulk goods storage
   facilities) 1841
     agricultural 2681
Hoppers (construction plant) (B4k)
Hoppers (elements of construction
   projects) (99.365)
     chute and hopper refuse
     disposal (52.13)
Horizontal dividing elements (2.9)
Horse racing facilities 564
Horses rearing and living facilities 4643
Horticultural facilities 264
Horticultural sundries (plant) (B8w)
Hose drying spaces 372
Hose pipes I

Hose reels (fire fighting services) (68.55)
Hospital school facilities 7172
HOSPITALS 41
Hostels 856
Hostility (U2)
Hot and cold water supply
   services (53.1)/(53.6)
   code (53) alone is usually used,
   particularly in project applications
Hot cupboards (culinary fittings) (73.5)
Hot storage facilities 966
Hot storage fittings (76.6)
   culinary display fittings (73.5)
Hot water central heating (56.4)
Hot water supply and space
   heating (56.4)
Hot water supply services (53.3)/(53.5)
   from common supply (53.3)
   from individual appliances (53.5)
   code (53.3) is used as the general
   position
Hot weather (H121)
Hot-dipped coatings u1
Hotels 852
Hothouse plants W
Hothouses 264
Hotplates (culinary fittings) (73.4)
House boats 87
Households (U1)
Housing Association housing 818
HOUSING FACILITIES 81
   for special classes of user 84
   historical 86
   individually designed 82
   temporary and mobile 87
Hovercraft stations 134
Hue (G5)
Human beings
   factors affecting works (L61)
   requirements (U3)
Human relations (Afk)
   project (or office)
   administration (A1fk)
Humidifiers (99.121)
   air treatment plant (57.5)
Humidity (L21)
Hunting facilities 565
Hurricanes (H122)
Huts, site (B1b)
Hydrants (fire fighting services) (68.55)
Hydrated lime q7
Hydration (L41)
Hydraulic binders q3
Hydraulic cement q2
Hydraulic energy (R1)
Hydraulic excavators (B2j)
Hydraulic lifts (66.1)
Hydraulics (J21)
Hydroelectric power stations 162
Hydrolysis (L5)
Hydromechanical energy (R1)
Hydrophones (B2t)
Hygiene (U46)
   activities in facilities (T194)
HYGIENE AND SANITARY FITTINGS
   as a whole (74)
   built-in or otherwise fixed (74)
   loose furniture and equipment (84)
Hygiene facilities 94
Hygroscopy (L23)
Hyperbolic paraboloid roofs (27.5)
Hypermarkets 344

I

Ice (H122)
Ice rinks 567
Identification (communications) (Agh)
Igneous stone e1
Ignitability (K44)
Ignition temperature (K44)
Illnesses (U461)
Illuminated signs (71.1)
   in external works (90.7)
Illumination (N7)
Illustrations facilities 765
Immersion heaters (99.123)
   hot water supply services (53.3)
Immersion vibrators (B6g)
Impact sound transmission (P3)
Impact strength (J4)
Impellers (99.33)
   piped and ducted services (59)
Import controls (A6j)
Impregnated fibres and felt n2
IMPREGNATION WORK V
Improvement (quality change) (X7)
   operations on existing works (W7)
Improvement planning areas 081
Impurity (L5)
IN SITU CASTING WORK E
In situ piling equipment (B2r)
Inadequate heat (M6)
Incandescence (N3)
Incandescent lighting sources (63.9)
Incentives (A5r)
Incinerators (99.11)
   refuse disposal (52.18)
   sanitary waste disposal (74.48)
Incombustibility (K41)
Income (A2s)
Indentation (J4)
Indented blockwork and brickwork F
Independent school facilities 718
Indexing (Ahi)
Indicators (99.4)
   electrical services (69)
   piped and ducted services (59)
Indirect costs (A2t)
Indoor plants W
Industrial activities (T12)
Industrial designers (Amw)
INDUSTRIAL FACILITIES 2
Industrial gases supply services (54.4)
Industrial relations (Afk)
Industrial towns 053
Industrial use (H7)
Industrialisation factors (E5)
Industrialised building systems (– –)
Inertia (R1)
Infants (U31)
Infants' school facilities 712
Infection (U466)
Infestation (L62)
Infill areas 085
Infill buildings 982
Infilling rigid sheet work R
Infillings (99.8)
Inflammability (K45)

Consult the schedules to check the context of items in the index. Italic entries are explained in the introduction.

Information (Ag)
  project (or office)
    administration (A1g)
Information activities (T176)
INFORMATION FACILITIES 76
Information preparation (Ah)
  designing and physical
    planning (A3h)
  project (or office)
    administration (A1h)
Infra red radiation (Q7)
Inland water transport facilities 133
Inlets (99.365)
  foam inlets (68.55)
Innovation (X8)
Inorganic materials z2
  fertilizers w6
  fibres m
In-patients facilities 418
Insect keeping facilities 4648
Insect resisting materials u3
Insecticides u3
  ancillary materials w9
Insects
  factors affecting works (L62)
  requirements (U51)
Insolation (M5)
INSPECTION (A7)
Inspection chambers (99.64)
  drainage services in general (52.9)
Inspection period (maintenance) (A8r)
Inspection pits (road vehicles) 127
Installing (D6)
  temporary services (D2)
Instructions, clients' (A3r)
Instrument engineering manufacturing
  facilities 2753
Instrument rooms (laboratory
  facilities) 732
Insulating coating work P
Insulating quiltwork K
Insulating rigid sheet work R
Insulation (M)/(Q)
  electrical (Q2)
  light (N2)
  sound (P2)
  thermal (M2)
Insurance (A9s)
Insurance facilities 335
Intake (electrical mains supply) (61)
Integrated financial/costs systems
  (A2) + (A4)
Integrated services (5.2)
Integration (X1)
Intense development areas 0831
Interchangeability (V7)
Intercity railways 111
Intercom services (audio
  communications) (64.32)
Interior design 997
Interlocking joints Z
Intermittent loaders (B2m)
Internal diameter (F47)
Internal drainage (52.6)
Internal environment (H6)
Internal spaces 997
Internal structure (matter) (E1)
Internal telephones (project or office
  administration) (A1s)
INTERNAL WALLS (22)
  secondary elements (32)
International legislative and
  administrative facilities 311
INTERNATIONAL PLANNING AREAS 02
Intersections (roads) 128
Interstitial condensation (L27)

Interview facilities 928
Interviews of personnel (A1mp)
Intruder detection and alarm
  services (68.2)
Invalids (U35)
Investigating equipment (B8q)
Investment facilities 337
Ionisation potential (Q4)
Iridescence (G5)
Iroko i3
Iron h1
Ironing facilities 953
Ironing fittings (75.4)
Ironmongery t6/t7
Irradiation (Q7)
Irrigation facilities 187
Islands 092
Islands (roads) 128
Isolation facilities (hospitals) 418
Isolation joints Z

J

Jacked pile foundations (17.3)
Jails 376
Jetties 138
Jib cranes, mobile (B3q)
Job centres 315
Job manuals
  project (or office)
    administration (A1jw)
Joining
  ease of joining (V6)
Joint fillers t4
JOINTING MATERIALS t
Jointless suspended ceilings (35.1)
JOINTS Z
  or use (99.9)
Joist hangers (99.76)
  for floors (23.9)
  for roofs (27.9)
  for floors and roofs (29)
Joist plus infill floors (23.4)
Joists (99.741)
  in floors (23.9)
  in roofs (27.9)
  in floors and roofs (29)
Judo facilities 562
Junction boxes (99.84)
  electrical services in general (69)
  lighting subcircuits (63.9)
  power subcircuits (62)
Junction pipes I
Junctions, road 128
Junior school facilities 712
Jury boxes 317
Jute j3

K

Karate facilities 562
Keene's cement r2
Keeps 86
Kelvin temperature (M7)
Kennels 4642
Kieselguhr p1
Kinetic energy (R1)
Kiosks 348
Kitchens 931
Kitchen suites (73.1)
  loose furniture and equipment (83)
Knotting v1
Knowledge (communications) (Ag)
Kraft paper j2

L

Labelling (Y8)
Laboratory facilities 732
Labour, construction (Amx)
  projects (or office)
    administration (A1mx)
Labour costs (A2t)
Labour exchange facilities 315
Labour protection plant (B1c)
Labour relations (Afk)
Lacquers v4
Ladder type boom trenchers (B2n)
Ladders (24.6)
Lagging (construction operations) (D6)
Lagging (elements) (99.534)
  *for air conditioning and ventilation
    services (57.9)*
  for liquids supply services (53.9)
  *for piped and ducted services in
    general (59)*
  for roofs (27.9)
  for space cooling services (55.9)
  for space heating services (56.9)
Lagging quilts K
Lake asphalt s1
Lake transport facilities 133
Lakes 093
Laminated glass o8
Laminates
  flexible sheets T
  rigid sheets R
Lamin board i4
Lamps (99.124)
  lighting services (63.9)
Land areas 092
  international or national in scale 02
Land drainage (11.5)
Land reclamation facilities 185
Land retention facilities 186
LAND TRANSPORT FACILITIES 11/12
LAND USE PLANNING AREAS 06
Landing pads 148
Landlords (U36)
Landscape, areas in 092
Landscape architects (Amw)
Landscape design 998
Landscape maintenance
  equipment (B8x)
Landscaped offices 32
Landscaped spaces
  external 997
  internal 998
Landscaping (in external works) (90.8)
Landslide protection facilities 1862
Landslides (H16)
Lantern lights (37.42)
Lapped joints Z
Lapping (D4)
Larch i2
Larders (facilities) 935
Larders (fittings) (73.5)
LARGE BLOCKS AND BLOCKWORK G
Large scale development areas 082
Laser beam facilities 156
Latches t7
  or use (99.953)
Latex binders r4
Lathes (B5j)
Laths H
Latrines 941
  site accommodation (B1b)
Lattices J
Launching ramps 2755
Launderettes 346
Laundries 952

Lav–Mai

Lavatories 941
Law
   see Legislation
Law courts 317
Law enforcement facilities 374
Lay lights (35.3)
Laybys 128
LAYING (D6)
Layout (F1)
Lead h8
Leaf fibres j3
Leakage (X2)
Lean-to buildings 982
Lean-to roofs (27.21)
Learned societies facilities 727
Learning activities (T17)
Learning facilities 729
   further education 728
   schools 718
Leather j6
Leather goods manufacturing
   facilities 2762
Lecture theatres 7291
Legal proceedings (Ajy)
Legations 316
Legislation (Ajk)
   designing and physical planning (A3j)
   financing (A2j)
   inspection and quality control (A7j)
   orders based on legislation (Ajm)
   project (or office) administration (A1j)
   regulations, bye-laws (Ajn)
Legislative facilities
   international 311
   national 312
   regional and local 313
Leisure facilities 5
Leisure gardens 587
Lending (finance) (A2u)
Lending facilities
   libraries 762
   parts of information facilities 768
Length (F47)
Lettering (71.1)
Levees 187
Level crossings 149
Levelling (D2)
LIBRARIES 76
Library practice (A1gm)
Licences (Ajo)
Licensed premises 517
Lidos 543
Life assurance for personnel (A1mr)
Life style (U2)
Lifeboat stations
   protective services facilities 371
   water transport facilities 136
Lift bridges 1825
Lift enclosures 9142
Lift lobbies 9142
Lift sludge pumps (B2t)
Lift wells (24.8)
LIFTING (D3)
   plant (B3p)
Lifts (construction plant) (B3w)
Lifts (transport services) (66.1)
LIGHT (N)
Light artificial granular aggregates p3
Light proofing (N2)
Lighthouses 136
Lighting fittings (63.9)
LIGHTING SERVICES (electrical) (63)
   in external works (90.6)
   temporary site services (B2c)
Lighting sources (luminaries) (99.124)
   lighting services (63.9)

Lightning (H122)
Lightning conductors (68.6)
Lightning protection services (68.6)
Lightweight aggregate concrete q7
   precast f5
Lightweight cellular concrete q6
   precast f4
Lime q1
   asbestos–silica–lime f6
   lime–cement q3
   lime–cement–aggregate mixes q4
Lime mortars and plasters q4
Limestone e2/e3
Line of balance (work study) (A5u)
Linear form planning 084
Linen j3
Lines J
Lines facilities, telecommunications 156
Lining coating work P
Lining rigid sheet work R
Linings (99.82)
   ceiling finishes (45)
   finishes to structure (49)
   roof finishes (47)
   wall finishes, internal (42)
Link buildings 9141
Linked buildings 982
Links (circulation facilities) 9141
Linoleum n4
Linseed oil v3
Lintels (99.741)
   in external wall openings (31.9)
   in internal wall openings (32.9)
   for project purposes, codes (31), (32)
   or (3–) are used as appropriate.
Liquid fuels w3
Liquid soap supply services (74.7)
Liquids (interactive factors) (L3)
Liquids (formless products) Y
Liquids storage facilities 964
LIQUIDS SUPPLY SERVICES (53)
Lithography (Ahq)
Live loads (J4)
Livestock facilities (agriculture) 265
   see also Animal welfare facilities
Living rooms 88
Load centre (electrical supply) (61)
Loadbearing internal walls (22.1)
Loadbearing wall structures (2.1)
Loadbearing walls (21.1)
Loaders
   continuous (B2k)
   intermittent (B2m)
Loading (D3)
Loading bays 125
Loading shovels (B2g)
Loads (energy consumption) (R3)
Loads (mechanics) (J4)
Loads, fire (K43)
Lobbies 913
   lift lobbies 9142
Local authority housing 818
Local government organizations (Akn)
   design and construction
   departments (Aks)
Local legislative and administrative
   facilities 314
Local offices of government
   departments 315
Local planning areas 036
Local shopping centres 342
Localised services (5.7)
Locate and fix operations (D6)
Lockers (storage fittings) (76.6)
Locks (securing devices) t7
   or use (99.951)

Locks (water transport facilities) 138
Locomotive sheds 117
Locomotives 118
   transport plant (B3k)
Lofts 993
Loggias 996
Long stay accommodation 448
Loop roads 1223
LOOSE FILL WORK C
LOOSE FILLS p
Loose fittings (as a whole) (8–)
LOOSE FURNITURE AND
   EQUIPMENT (8–)
   in external works (90.7)
Loose paving work C
Lorries (B3d)
Loudness (P6)
Loudspeakers (99.125)
   communications services (64.9)
Lounges 922
Louvre windows (31.43)
Louvred openings (31.8)
Louvred suspended ceilings (35.2)
Low heat cement q2
Low income people (U36)
Low rise buildings 981
Low temperature (M7)
Low voltage electrical mains
   supply (61.6)
Lubricating facilities 127
Luggage rooms 961
Luminaires (lighting sources) (99.124)
   lighting services (63.9)
Luminance (N5)
Luminescence (N3)
Luminosity (N5)
Luminous intensity (N7)
Lustre (G5)

M

Macadam s5
Macadam plant
   manufacturing plant (B4f)
   spreaders (B6c)
Machine bases (33.2)
Machine rooms 285
Machines (requirements) (U53)
Machines as heat source (56.16)
Machines as power source (5.16)
Mack's cement r2
Magistrates' courts 317
Magnesia r3
   preformed f8
Magnesite based refractory
   materials g6
Magnesium oxychloride r3
Magnesium oxysulphate r3
Magnetism (Q6)
Mahogany i3
Mail order stores 341
Mail rooms 157
Main pipes l
Main roads 1221
Mains (electrical supply) (61)
   public supply (61.6)

Consult the schedules to check the context of items in the index. Italic entries are explained in the introduction.

Mai–Mod

Maintenance (W2)
    activities in facilities (T195)
    administration of handover (A8r)
    landscape equipment (B8x)
MAINTENANCE AND CLEANING FITTINGS
    as a whole (75)
    built-in or otherwise fixed (75)
    loose furniture and equipment (85)
MAINTENANCE FACILITIES 95
Maisonettes 817
MAKING GOOD (D7)
    administration of handover (A8r)
Maladjusted people
    see Handicapped people
Males (U33)
Malleable iron h1
MALLEABLE SHEETS AND SHEET WORK M
Mammals (requirements) (U51)
Management
    activities (T131)/(T132)
    in construction industry (A)
Management resources (U6)
Manholes (99.64)
    drainage services (52.9)
Man-made features of the environment 998
Man-made resources (U6)
Manoeuvrability (V3)
Mansard roofs (27.28)
Manual operating devices t7
    or use (99.963)
Manual refuse disposal (52.14)
Manual workers (U37)
Manually operated fire fighting services (68.55)
Manuals
    building owner's (A8h)
    office (A1jw)
Manufactured materials z3
Manufacturers (Amu)
    project (or office) administration (A1mu)
Manufacturing
    activities in facilities (T127)/(T128)
    controls (A5j)
    plant (B4)
    production factors (E3)
    resources (U6)
MANUFACTURING FACILITIES 27
Manures w6
Maps, site investigation and surveying (A3s)
Marble e2
Marinas 546
Marine engineering facilities 2755
Marital status of people (U34)
Maritime water transport facilities 131
Market gardening facilities 264
Market research (A6o)
Market towns 053
Marketing (A6f)
Markets, shopping 342
Married people (U34)
Marshalling yards 116
Marshes 092
Martin's cement r2
Masonry F
Masonry cement q2
Mass concrete q4
Mass retaining walls (16.2)
Masses (unorganised groups) (U1)
Mastic asphalt s4
Mastic asphalt manufacturing plant (B4e)
Mastics t4

Masts 981
MATERIALS a
Materials costs (A2t)
Materials found on site (A9t)
Materials handling (D3)
Materials handling (movement) facilities 915
Materials protection plant (B1d)
Maternity and child clinics 422
Maternity hospitals 415
Matt (reflecting properties) (N41)
MATTER (L)
Maturing (X6)
Mausoleums 67
Maximum security prisons 3762
Mayors' parlours 314
Means of escape (K24)
Measurement of quantities (A4s)
Measurement of sound (P7)
Measuring devices (99.4)
    electrical services (69)
    piped and ducted services (59)
Mechanical circulation services (66)
Mechanical energy (R1)
Mechanical engineering manufacturing facilities 2752
Mechanical handling services (transport/storage) (66.8)
Mechanical power supply facilities 163
MECHANICS (J)
Mechanization (information processing) (Agn)
    cost planning (A4gn)
    production planning (A5gn)
Media (communication) (Agi)
Media (paints) v3
Medical gases supply services (54.4)
Medical research facilities 427
Medical treatment
    activities (T141)/(T142)
Medical treatment facilities 41/42
Medium board j1
Medium rise buildings 981
Medium voltage electrical mains supply (61.6)
MEETING HOUSES 64
Meetings (Agj)
    buying (A6gj)
    designing and physical planning (A3gj)
    production planning (A5gj)
    project (or office) administration (A1gj)
Melamine resins n6
Membrane structures (2.6)
Membrane work L
Memoranda (Ajm)
Memorials 984
Men (U33)
Mental hospitals 413
Mentally handicapped people (U35)
Mentally handicapped people's facilities
    common facilities 988
    school facilities 7172
    welfare facilities 444
Merchants (bulk materials) 348
MESHES AND MESH WORK J
METAL h
Metal aggregate granolithic q5
Metal backed paper j2
Metal faced wood laminates i4
Metal manufacturing facilities 2751
Metallic coatings u1
Metallic paints v9
Meteorological factors (H12)
Meteorological stations 737

Meters (99.421)
    communications services (64.9)
    electrical services in general (69)
    electrical supply services (61.9)
    gases supply services (54.9)
    liquids supply services (53.9)
    piped and ducted services in general (59)
Method study (A1p)
    production planning (A5p)
Methodology
    cost planning (A4f)
    designing and physical planning (A3f)
Methods of measurement (quantity surveying) (A4s)
Metric system (F7)
Metropolises 052
Metropolitan counties planning areas 033
Metropolitan districts planning areas 035
Mezzanines 993
Mica, expanded p3
Microbes (L63)
Microclimate (H121)
Microfilming (A1t)
Microphones (99.125)
    communications services (64.9)
Micro-wave beam facilities 156
Micro-wave radiation (Q7)
Middle school facilities 712
Mild steel h2
Milk bars 515
Mill board j2
Mills, heavy industry 281
Mineral extraction facilities 164/167
Mineral supply facilities 163/168
Mineral waste disposal facilities 176
Mineral wool m1
Mines
    solid fuels 166
    other minerals 167
Ministries (Akn)
Ministries facilities 312
Minor roads 1222
Minors (U31)
Mirror reflecting properties (N41)
Mirrored glass o7
Mirrors (sanitary and hygiene fittings) (74.8)
Miscibility (L37)
Missile sites 3758
MISSION HALLS 64
Mistakes in use (T7)
Mis-use (T7)
Mitred joints Z
Mix proportions (E1)
Mixed commercial developments 331
Mixed hydraulic binders q3
Mixed natural/synthetic fibres n9
Mixes (interactive factors) (L3)
Mixes (formless products) Y
Mixing (D4)
Mixing plant, concrete (B4g)
Mixing valves (99.411)
    liquids supply services (53.9)
Mixtures (L3)
Mobile buildings 983
Mobile jib cranes (B3q)
MOBILE RESIDENTIAL FACILITIES 87
Models (Z(08))
Modern monuments 986
Modernisation (quality change) (X7)
    operations on existing works (W7)
Modification (W6)
Modifying agents u2
Modular co-ordination (F43)

Moisture barriers (99.532)
   external walls (21.9)
   roofs (27.9)
   structure generally (29)
Moisture content (L31)
Moisture proofing (L34)
Molar earth p1
Moles (coastal protection) 1863
Momentum (R1)
Monasteries 66
Monitor lights (37.42)
Monitoring services (68.7)
Monitoring stations
   (telecommunications) 156
Monolithic floors (23.2)
Monolithic internal walls (22.6)
Monorails 113
   transport plant (B3k)
Monuments 984
Moorings 138
Moorland 092
Morals (U2)
Morgues 177
Mortar machines (B4i)
Mortars
   clay and refractory r1
   lime and cement q4
   synthetic resin r4
Mortice and tenon joints Z
Morticing (D5)
Mortuaries 177
Mosaic work S
MOSQUES 65
Moss w6
Motels 853
Mothers (U32)
Motor cycle racing facilities 564
Motor racing facilities 564
Motor roads 121/122
Motors (construction
   plant) (B8d)
Motors (elements) (99.127)
   transport services (66.9)
Motorways 121
Mould (L64)
Mouldings (99.98)
   finishes to structure (49)
Moulds (formwork) (B2d)
Mounds (ground relief) (11.1)
Mountain railways 113
Mountainous areas 092
Movability (V3)
Movable bridges 1825
Movement (X3)
   dimensional change (F47)
   circulation activities (T191)
   forces opposing motion (J5)
Movement facilities 914
Movement joints Z
Moving and guiding devices (99.3)
Moving pavements (66.5)
Mowing (D4)
Multiple glazing o5
Multi-storey buildings 981
Murals 984
Murals display facilities 757
Museums 756
Music academies 724
Music practice rooms 522
Musical facilities 522

N

Nailability (V6)
Nailing tools (B5d)
Nails t6
National legislative and administrative
   facilities 312
National libraries 761
National parks 0871
National planning areas 02
Nationalised industries' offices 32
Nations (U1)
Natural adhesives t3
Natural aggregates p1
Natural areas 091/095
   in the service of man 087
   international or national in scale 02
Natural environmental factors (H1)
Natural features of the environment 998
   areas in the service of man 087
Natural fills p1
Natural gas supply (54.1)
Natural light (N11)
Natural resources (U6)
Natural rubber n5
NATURAL STONE e
Naturally occurring materials z3
Navy facilities 3752
Needy people (U36)
Negotiation
   buying (A6f)
   project (or office) administration (A1f)
Neighbourhoods 058
Neoprene n5
   cement – neoprene latex r4
Nervous system (U41)
Netball facilities 564
Netting J
Network analysis (A5t)
Neurosurgical facilities in hospitals 4172
New towns 054
Nickel h9
Nickel bronze h6
Nickel plated coatings u1
Nickel silver h6
Night clubs 537
Nitriding treatments u5
Nitrile n5
Noise (P1)
Noise attenuators (99.535)
   *air conditioning and ventilating
   services (57.9)*
   sound control services (68.8)
Nomenclature (Agh)
Nominations (contracts) (A4u)
Non-combustibility (K41)
Non-ferrous metals h4/h9
Non-hydraulic binders q3
Non-loadbearing internal walls (22.3)
Non-loadbearing walls (21.3)
Non-metallic coatings u1
Non-powered tools (B7)
Non-reflection (light) (N41)
Non-structural section work H
Non-thrust resistant wall storage
   structures 1843
   agricultural 2683
Non-woody plants W
Northlight roofs (27.8)
Northlights (37.42)
Nosings (99.87)
   stair nosings (44)
Notice boards (71.1)
   in external works (90.7)
Notices, warning (B1c)

Nuclear fission waste products disposal
   facilities 176
Nuclear power stations 162
Nuclear radiation (Q7)
Nunneries 66
Nurseries 446
Nurseries (horticultural) 264
Nursery school facilities 711
Nursing facilities 418
Nursing homes 442
Nutrition (U463)
Nuts and bolts t6
Nylon n6

O

O and M
   project (or office)
   administration (A1p)
Oak i3
Obelisks 984
Objectives (Afg)
   project (or office)
   administration (A1fg)
Observatories 737
Obsolescence (T4)
Occupancy (H7)
Occupants as heat source (56.16)
Occupants as power source (5.16)
Occupational therapy facilities in
   hospitals 4175
Occupations (U37)
Occupiers (U36)
Ocean defence facilities 1863
Ocean transport facilities 131
Odour (G8)
Office manuals (A1jw)
Office organization (A1)
OFFICES 32
   commercial facilities 33
   housing with offices 818
   official administrative facilities 31
   records, archives, patent offices 767
   site offices (B1b)
   sorting offices 157
OFFICIAL ADMINISTRATIVE
   FACILITIES 31
Official advice (Ajp)
Official representation facilities 316
Oil as power source (5.12)
Oil extraction and storage facilities 164
Oil heating (56.12)
Oil paints v5
Oil supply facilities 164
   common facilities 971
Oil supply services (53.7)
Oil well cement q2
Oils v3
Old people (U32)
Old people's clinics 422
Old people's homes 447
Old people's hospitals 416
Old people's housing 843
One and a half brick brickwork F
One brick brickwork F
One-off housing units 82

**Consult the schedules to check
the context of items in the index.
Italic entries are explained in
the introduction.**

'One way' glass o7
On-site assembly (E4)
Opacity (G6)
Opal glass o3
Opaque glass o3
Open air swimming pools 543
Open plan spaces 997
Open prisons 3761
Open spaces in built-up areas 058
Open spaces, recreational 06:5
Open well stairs (24.5)
Open-cast workings
    solid fuel 166
    other minerals 167
Opening devices t7
    or use (99.961)
Openings completion (3–)
Opera houses 523
Operating devices t7
    or use (99.96)
Operating theatres 4172
OPERATION FACTORS (W)
Operational research (A1o)
OPERATIONS, CONSTRUCTION (D)
Opinion, public (U2)
Optical factors (N)
Ordering
    buying for construction projects (A6r)
    supply factors (Y8)
Organic materials z2
    fertilizers and manures w6
    solvent preservatives u3
Organized groups (U1)
ORGANIZATION, PROJECT AND OFFICE (A1)
Organization and methods
    project (or office) administration (A1p)
Organizations (U1)
    construction industry (Ak)
Organizing information (Ag)
Ornamental buildings 984
Ornamental grasses W
Ornamental shrubs W
Orphanages 446
Outdoor activities (T199)
Outdoor fittings (90.7)
Outdoor recreation facilities 587
Outdoor sports facilities 568
Outgoings, financial (A2t)
Outlets (99.365)
    gases supply services (54.9)
Outline proposals (A3)
Outpatients facilities 418
Ovens (culinary fittings) (73.4)
Overcrowding (H7)
Overhauling (W5)
Overhead railways 111
Overheating (M6)
OVERLAPPING RIGID SHEET WORK N
Overlays (99.84)
Overload (energy consumption) (R3)
Overnight accomodation (welfare facilities) 448
Overshadowing (H2)
Overspill areas 058
Owners' manuals (A8h)
Ownership (U2)
    characteristics of areas 096
Oxide film formation coatings u1

P

PERT (A5t)
PSALI (N13)
PVA n6
    cement–PVA emulsions r4
PVC n6
    cement–PVC emulsions r4
Package deal services (Y9)
Package dealers (Aky)
Packaging (Y8)
Pad foundations (16.4)
Paediatric hospitals 416
Paging services (audio communications) (64.13)
Painting equipment (B5h)
Paintings (fittings) (78.6)
    loose fittings (88)
Paintings display facilities 757
PAINTS v
    bitumen paints s1
PAINTWORK, DECORATIVE V
Palaces 316
    episcopal 61
Palletized transport/storage services (66.8)
Panel and slab structures (2.3)
Panelled suspended ceilings (35.1)
Panels R
    large panels and panel work G
Panels (temporary works) (B2d)
Panes R
Pantries 935
Paper industries facilities 2773
Papers and papering work T
    proofing papers L
    vegetable materials j2
Parades 982
Parapet walls (21.8)
    roof secondary elements (37.6)
Parasites (U466)
Parcels offices 157
Parents (U32)
Parian cement r2
Parking facilities 125
    common facilities 963
Parks 587
    national parks 0871
Parliaments 312
    European 311
Parquet work S
Particles Y
PARTITIONS (22)
    secondary elements (32)
    folding partitions (32.54)
    sliding partitions (32.54)
Partners (A1my)
Partnerships (Akv)
    project (or office) administration (A1mq)
Party floors (23.8)
Party walls (22.8)
Passages 9141
Passenger lifts (66.1)
Passengers (U37)
Passing places (roads) 128
Pasteboard j2
Pastes Y
Pastoral centres 61
Patents (Ajx)
Patents offices 767
Paternosters (66.1)
Pathology facilities in hospitals 4174
Paths 1234
    in external works (90.41)
Patio doors/windows (31.3)

Patios 998
Pattern (G4)
Pattern staining (L44)
Patterned glass o2
Pavement lights (37.47)
Pavements (hard floor beds) (13.1)
    in external works (90.41)
Pavements (road transport facilities) 128
Pavilions 996
    sports 568
Paving equipment (B6b)
    concrete (B6d)
Paving work
    blocks, bricks, stone F
    loose C
    slabs (rigid tiles) S
    thick coatings P
Payment methods (Y3)
Peace (U2)
Peak load (energy consumption) (R3)
Peat bog 092
Peat cuttings 166
Pebbledash work P
Pebbly soils p1
Pedestrian facilities 988
Pedestrian streets 1231
Pendant lighting fittings (63.9)
Penetration (L41)
Penitentiaries 376
Pens (agricultural enclosures) 2688
Pensions
    costs (A2t)
    project (or office) administration (A1mr)
Pentathlon facilities 568
People
    factors affecting works (L61)
    requirements (U3)
Perennials W
Perforated bricks F
Performance factors (J)/(T)
    suitability (T3)
Performing arts facilities 52
Perimeter (F47)
PERIPHERAL SUBJECTS (Z)
Perlite, expanded p3
Permanent way (railways) 118
Permeability (L5)
    gas (L26)
    liquid (L34)
    solids (L42)
Personal details of personnel (A1mr)
PERSONNEL (Am)
    project (or office) administration (A1m)
Persons
    factors affecting works (L61)
    requirements (U3)
Petrol filling stations 126
Petrol supply services (53.7)
Petroleum bitumen s1
Petroleum extraction facilities 164
Petroleum waste disposal (52.4)
Pharmacies in hospitals 4177
Phenolic resins n6
Phosphated coatings u1
Phosphor bronze h6
Photocopying (A1t)
Photograph libraries 765
Photography (Ahk)
    project (or office) administration (A1t)
Photography facilities 32
    common facilities 926
PHYSICAL ENVIRONMENT 0/9
Physical medicine facilities in hospitals 4175

Phy–Pre

PHYSICAL PLANNING (A3)
Physical training facilities 562
Physically handicapped people (U35)
Physically handicapped people's facilities
    common facilities 988
    housing facilities 848
    school facilities 7173
    welfare facilities 445
Physiographic factors (H16)
Physiological factors (U41)
Physiotherapy facilities in hospitals 4178
Piano trolleys (B3i)
Pier buildings 988
Piers 138
Pigments v2
Pigs rearing and living facilities 4646
Pile caps (17.9)
Pile driving equipment (B2r)
PILE FOUNDATIONS (17)
    temporary sheet piling (B2r)
Pile frames (B2r)
Pillars (99.751)
Pine i2
Pin-jointed building frames (28.7)
Pins H
Pipe bends I
Pipe bridges 1824
Piped services (5–)
Piped supply of solids by pressurisation or suction (54.8)
Pipelines 181
    gas 165
    oil 164
Pipes and pipe fittings
    (elements) (99.361)
    drainage services (52.9)
    liquids supply services (53.9)
    gases supply services (54.9)
    piped and ducted services in general (59)
    space cooling services (55.9)
    space heating services (56.9)
PIPES AND PIPEWORK (FORMS, CONSTRUCTIONS) I
Pise de terre g1
Pitch s1
Pitch fibre n2
Pitch mastic s4
Pitched roofs (27.2)
Pitches 568
Pits
    agricultural facilities 2688
    road vehicle inspection pits 127
    solid fuel mines 166
PLACE (Z)
PLACING (D6)
    ease of placing (V6)
PLACING PLANT (B6)
    concrete (B4g)
Plain casting work E
Plain concrete q4
Plain glass o1
Plain tiles and tiling work N
Plan of work
    production planning (A5f)
    project (or office) administration (A1fx)
Planetariums 756
Planks H
Planners (Amw)
    central and local government departments (Aks)

Planning
    cost (A4)
    designing and physical planning (A3)
    production (A5)
    see also Town and country planning
PLANNING AREAS O
Plans
    for planning areas 038
    layouts (F1)
    objectives (Afg)
PLANT (B)
    costs (A2t)
Plant containers (71.2)
    in external works (90.7)
Plant rooms 97
Planted and unplanted beds (13.4)
    in external works (90.42)
PLANTS W
    factors affecting works (L64)
    flora 0874
    requirements (U52)
    see also Vegetation
Plaster of Paris r2
Plastering machines (B6j)
Plasters
    cement, lime q4
    gypsum r2
Plasticisers u2
Plasticity (J7)
Plastics n6/n9
    binders and mortars r4
Plastics faced wood laminates i4
Plated coatings u1
Plates (rigid sheets) R
Plates, vibrating (B6f)
Platform roofs (27.1)
Platform trucks (B3h)
Platforms (71.3)
    loose furniture and equipment (81)
Platforms, railway 114
Play activities (T158)
Play facilities 928
Play fittings (77)
    in external works (fixed and loose) (90.7)
    loose furniture and equipment (87)
Playgrounds 585
Playing fields 568
Plugs (for fixings) t6
Plywood i4
Pneumatic conveyors (66.5)
Pneumatic energy (R1)
Pneumatic message systems (64.8)
Pneumatic-tyred tower cranes (B3t)
Pneumatic-tyred tractor shovels (B2m)
Polar factors (H11)
Polarisation (light) (N42)
Polders 185
Poles (vertical constructions) 981
Police stations 374
Policy (Afg)
    project (or office) administration (A1fg)
Polio handicapped people
    see Physically handicapped people
Polishes u5
Pollution (L7)
Polo facilities 565
Polyamides n6
Polybutadiene n5
Polychloroprene n5
Polyesters n6
Polyethylene n6
    chlorosulphonated polyethylene n5
Polyisobutylene n5
Polyisoprene n5
Polypropylene n6

Polystyrene n6
Polysulphides n5
Polytechnics 722
Polythene n6
    bitumen polythene n2
Polyurethanes n6
Polyvinyl acetate n6
    cement–PVA emulsions r4
Polyvinyl chloride n6
    cement–PVC emulsions r4
Ponds 093
Pools 093
Pools (elements) (13.6)
    in external works (90.43)
Pools, swimming 541/544
Poor people (U36)
Population (U2)
    areas by type of population 097
Porcelain g3
Porches 913
Porches (elements) (31)
Portable fire fighting systems (68.56)
Portal frame bridges 1828
Portal frames (28.2)
Portland cement q2
Ports 134
Post offices 157
Postal communications facilities 157
Postal procedures
    project (or office) administration (A1s)
Postgraduate teaching hospitals 411
Posture (U41)
Potato peelers (73.4)
    loose furniture and equipment (83)
Potential energy (R2)
Pots (hollow blocks) F
Pottery industries facilities 2771
Poultry rearing and living facilities 4647
Pouring (D6)
Poverty (U2)
Powder p6
Power capacity (R3)
Power plant (B8b)
Power rammers (B6f)
POWER SERVICES (62)
Power sources for heating (56.1)
Power sources for services (5.1)
Power stations 162
POWER SUPPLY FACILITIES 16
    common facilities 971
Practice facilities
    entertainment 528
    music 522
    sports 568
Preaching facilities 68
Precast concrete f1/f5
PRECAST MATERIALS WITH BINDER f
Precasting work E
Precincts 9141
Prefabricated building systems (– –)
Prefabrication (E3)
PRELIMINARIES AND GENERAL CONDITIONS A
    used in specifications applications
Premises
    project (or office) administration (A1r)

Consult the schedules to check the context of items in the index. Italic entries are explained in the introduction.

Pre–Rai

PREPARATION OF SITE (D2)
    plant (B2)
Preparatory school facilities 712
PRESENTATION, FORMS OF (Z)
Presenting information (Ag)
Preservation (X8)
Preservatives u3
Presidential residences 316
Pressed bricks F
Pressurised piped supply of solids (54.8)
Pre-stressed casting work E
Pre-stressed concrete q4
Pre-stressing equipment (concrete manufacture) (B4h)
Preventative medicine factors (U46)
Price evaluation (A4t)
Prices (Y2)
Pricing (A4t)
Pricking-up work P
PRIMARY ELEMENTS OF STRUCTURE (2–)
Primary roads 1221
    motorways, autobahns 121
Primary school facilities 712
Prime movers (B3d)
Primers v1
    complete emulsion paint systems v6
    complete oil paint systems v5
Priming coating work V
Printing (Ahq)
    costs (A2t)
Printing industries facilities 2773
Priories 66
Prisons 376
Private design and construction organizations (Akv)
Private housing 818
    see also Houses
Private patients' hospital facilities 418
Process control services (68.7)
PROCESS MODIFYING MATERIALS u
PROCESSES (E)/(Y)
Processing activities (T197)
Processing devices (99.1)
Processing facilities 97
Produce
    animal (U51)
    plant (U52)
Produce facilities 264
PRODUCT FORMS AND CONSTRUCTIONS A/Z
Production facilities
    heat supply 161
    industrial facilities 2
    power supply 162
Production factors (E2)
Production information (A3)
PRODUCTION PLANNING (A5)
PRODUCTS Y
Professional bodies (Akp)
Professional handbooks
    project (or office) administration (A1jw)
Professional offices 32
Professional workers (U37)
Professions (Am)
Profiled sheets and sheet work N
Profiles (sections) H
Profit sharing
    project (or office) administration (A1mr)
Profits (A2r)
Programming techniques (A5s)
Progress control (A5f)
PROJECT ORGANIZATION (A1)
Projecting blockwork and brickwork F

Projection facilities 768
    Cinemas 525
PROJECTS (construction sites as a whole) (– –)
Prominences (ground relief) (11.1)
Proofing
    biological change (L6)
    chemical change (L5)
    contamination (L7)
    deterioration (L8)
    gas (L26)
    heat (M2)
    light (N2)
    liquid (L34)
    sound (P2)
    weather (H123)
PROOFING SHEETS AND SHEET WORK L
Propane gas supply (54.1)
PROPERTIES (E)/(Y)
PROPERTY MODIFYING MATERIALS u
Property spaces (drama) 524
Proportion (G1)
Props (temporary works) (B2d)
Protection (X8)
    electrical (Q2)
    electromagnetic radiation (Q7)
    fire (K2)
    heat (M2)
    light (N2)
    sound (P2)
    vibration (J6)
Protection and control services (68)
PROTECTION OF SITE (D1)
    plant (B1)
Protective clothing (B1c)
Protective coating work
    film coatings V
    thick coatings P
Protective malleable sheets and strips M
PROTECTIVE MATERIALS u
Protective service activities (T137)
Protective service facilities 37
Pruning (D4)
Psychiatric hospitals 413
Psychology, social (U2)
Public address services (audio communications) (64.33)
Public design and construction organizations (Aks)
Public enquiries (Ajy)
Public health engineering 17
Public houses 517
Public lavatories 941
Public library facilities 762
Public opinion (U2)
PUBLIC RELATIONS (Ai)
    project (or office) administration (A1i)
PUBLICITY (Ai)
    project (or office) administration (A1i)
Publishing (Ahp)
Publishing industries facilities 2773
Pulleys (99.34)
    construction plant (B3x)
    transport services (66.9)
Pulpboard j2
Pulverised fuel ash p4
    sintered pulverised fuel ash p3
Pumping D3
Pumping stations 1713

Pumps (elements) (99.33)
    drainage services (52.9)
    hot water central heating (56.4)
    liquids supply services (53.9)
    piped and ducted services in general (59)
    space cooling services (55.9)
    space heating services (56.9)
Pumps (plant) (B2t)
Purchase price (Y3)
Purchases
    costs of purchases (A2t)
Purification (quality change) (X7)
    factors affecting works (L7)
Purlins (27.9)
Push dozers (B2g)
Putty t4
    lime putty q1
Puttying equipment (B5h)

Q

Quadrangles 998
Quality change (X7)
QUALITY CONTROL (A7)
Quantifying (A4s)
Quantity surveyors (Amw)
Quarantine facilities 468
Quarries 167
Quartz aggregate granolithic q5
Quays 138
Quietness (P6)
QUILTS AND QUILTWORK K
Quotations (estimates) (A4t)

R

RIBA Plan of Work (A1fx)
Races (ethnic groups) (U1)
Racially mixed areas 097
Racing facilities 564
Rack railways 113
Racking (storage fittings) (76.4)
    loose furniture and equipment (86)
Radar facilities 146
Radial electrical distribution (61.1)
Radial roads 1223
Radiation
    electromagnetic (Q7)
    thermal (M35)
Radiation shelters 3757
Radiators (99.123)
    hot water central heating (56.4)
    space heating services in general (56.9)
Radio facilities 152
Radio services (audio communications) (64.31)
Radio therapy facilities in hospitals 4178
Radio wave facilities 156
Radio-active wastes disposal (52.4)
Radioactivity (Q7)
Radio–diagnostic facilities in hospitals 4171
Radius (F47)
Raft foundations (16.4)
Rafters (27.9)
Rail mounted tower cranes (B3t)
RAIL TRANSPORT FACILITIES 11
Rail vehicles
    manufacturing industry facilities 2756
    rail transport facilities 118
Railroads (plant) (B3k)
Rails (site protection) (B1a)

Rai-Rev

Railway engineering 11
Rain (H122)
Rainfall
   areas by rainfall 095
Rainwater drainage (52.5)
Rainwater goods (52.5)
Raked joints Z
Raking bond blockwork and brickwork F
Rammers (B6f)
Ramps (24.7)
Ramps, launching 2755
RAMPS AND STAIRS (carcass alone, or including soffites) (24)
   secondary elements (34)
   finishes (44)
Ranges, shooting
   defence 3755
   sports 565
Rapid hardening cement q2
Rates (A2t)
RATIONALIZATION (Ap)
Rayons n6
Reactivity (chemical) (L5)
Reading facilities 928
Rebated joints Z
Receivers (99.135)
   communications services (64.9)
Receiving stations (telecommunications) 156
Reception spaces 913
Recessed lighting fittings (63.9)
   as part of suspended ceilings (35.9)
Reclamation (X8)
Reclamation facilities 185
Reconstructed stone f3
Reconstruction (W7)
Record offices 767
Recorded communications (Agk)
Recording devices (99.4)
   electrical services (69)
   piped and ducted services (59)
Recording information (Ag)
Recording services (audio communications) (64.34)
Recording stations (scientific facilities) 737
Records, personnel (A1mr)
Recovery (X8)
Recovery rooms 922
Recreational activities (T15)
RECREATIONAL FACILITIES 5
Recreational open space 06:5
Recreational use (H7)
Rectangular hollow sections H
Re-decorating (W2)
Redevelopment areas 082
Reduction (chemical) (L5)
Redwood i2
Reeds j3
Refectories 511
Reference library facilities 768
Refineries, oil 164
Reflecting glass o2
Reflection factors
   light (N41)
   sound (P4)
Reformatories 3765
Refraction (light) (N43)
Refractory materials g6
   refractory mortar r1
Refractory temperature (M7)
REFRESHMENT FACILITIES 51
Refrigerated storage facilities 965
Refrigerated storage fittings (76.6)
   culinary display fittings (73.5)
Refrigeration plant (B8j)

Refrigeration services (55)
   from central plant (55.1)
   from local plant (55.5)
Refrigerators (culinary fittings) (73.5)
Refuges (roads) 128
Refuse disposal facilities 175
Refuse disposal services (52.1)
Regional legislative and administrative facilities 314
Regional factors (H11)
Regional planning areas 031
Regulations (Ajn)
   building regulations (A3j)
Rehabilitation (quality change) (X4)
   operations on existing works (W7)
Rehearsal facilities 528
Reinforced brickwork F
Reinforced casting work E
Reinforced concrete q4
Reinforced plastics n8
Reinforcement (99.79)
   to building frame (beams, columns, slabs) (28.9)
   to floors (23.9)
   to roofs (27.9)
   to structure generally (29)
   to walls (21.9)
Reinforcement (quality change) (X7)
Reinforcing materials u9
Relative humidity (L21)
Relaxation facilities 922
Relaying services (audio communications) (64.33)
Relays (99.416)
   communications services (64.9)
Release agents (B2d)
Relics spaces 67
RELIGIOUS FACILITIES 6
Religious use (H7)
Reliquaries 67
Remote control devices t7
   or use (99.962)
Removal (X2)
Rendering work P
Renewal (quality change) (X4)
   operations on existing works (W7)
Renewal areas 0832
Renovation (W7)
Rentals (A2t)
Repair facilities
   aircraft 147
   boats (water transport) 137
   boats (manufacturing facilities) 2755
   common transport facilities 149
   industrial 288
   rail vehicles 117
   road vehicles 127
Repairing (W5)
   construction operations (D7)
Repellants, water u6
Replacement (W7)
Replacement pile foundations (17.2)
Representatives, sales
   visits (A6gj)
Reprographic facilities 766
Reptiles (requirements) (U51)
REQUIREMENTS (E)/(Y)
   designing and physical planning (A3r)
RESEARCH (Ao)
   education and training (An)
   information handling (Ag)
   market research (A6o)
   operational research (A1o)
Research activities (T173)/(T174)

Research facilities 731
   medical 427
Reservoirs 1711
Residential activities (T18)
RESIDENTIAL FACILITIES 8
   clubs 536
   religious facilities 66
   school facilities 7178
Residential use (H7)
Resins v4
   oil-resin paints, synthetic resin paints v5
   synthetic resin-based emulsions v6
   synthetic resin compounds r4
   synthetic resins n5/n6
Resistance (X8)
   deformation (J3)
   electrical (Q3)
   fire (K3)
   forces opposing motion (J5)
   sound (P3)
   thermal (M31)
Resistivity (M31)
Resistors (99.362)
   electrical services (69)
Resonance (P3)
Resource centres (information facilities) 765
Resources (U6)
   consumption (T6)
   financial (A2)
   production planning (A5f)
Respiratory system (U41)
Responsibility (Afk)
   project (or office) administration (A1fn)
Rest and work facilities 921
REST AND WORK FITTINGS
   as a whole (72)
   built-in or otherwise fixed (72)
   loose furniture and equipment (82)
Rest facilities 921
Rest fittings (72.1)
   loose furniture and equipment (82)
Rest rooms 922
Restaurants 512
Resting (activities) (T192)
Restoration (quality change) (X8)
   operations on existing works (W7)
Restoration planning areas 081
Retaining walls (16.2)
RETAINING WALLS AND FOUNDATIONS (16)
Retardant treatment (fire) (K21)
Retardants u2
Retarded hemihydrate r2
Retarded people
   see Handicapped people
Retractable roofs (27.8)
Retraining (An)
Retreats 66
Retrieval of information (Agm)
   project (or office) administration (A1gm)
Retrochoirs 68
Re-use (T5)
Reverberation (P3)
Revetments 1864

Consult the schedules to check the context of items in the index. Italic entries are explained in the introduction.

Rev – Sec

Revolutionary change (X4)
Revolving doors (31.53)
    in internal openings (32.53)
Rhythm (X3)
Ribbed sheets
    flexible T
    rigid R
Ribbon form planning 084
Rich people (U36)
Riding school facilities 565
RIGID SHEETS AND SHEET OVERLAP WORK N
RIGID SHEETS AND SHEET WORK R
RIGID TILES AND TILE WORK S
Rigid-jointed building frames (28.7)
Rigs
    gas 165
    oil 164
Ring main electrical distribution (61.2)
Ring roads 1223
Rinks, ice 567
Rippers (B2g)
Rising damp (L31)
Rising main electrical distribution (61.3)
Rising pipes I
Rising walls (footings) (16.4)
Risks (U47)
River transport facilities 133
Rivers 093
Riverside areas 092
Riveted joints Z
Riveting (D4)
    tools (B5d)
Rivets t6
Road haulage systems 12
ROAD TRANSPORT FACILITIES 12
Road vehicles
    manufacturing facilities 2756
    road transport facilities 128
Roads 12
    built-up areas 058
    external works (90.41)
    site roads (B2c)
Robing rooms 318
Rock
    see Stone
Rock blasting equipment (B2b)
Rock drilling equipment (B5c)
Rock plants and rock work W
Rodding eyes (99.64)
    drainage services (52.9)
Rodents rearing and living facilities 4648
Rods H
ROLES (PERSONNEL) (Am)
    project (or office)
    administration (A1ms)
Rolled asphalt s5
Rolled asphalt manufacturing plant (B4f)
Roller bridges 1825
Roller conveyors (B3j)
Roller grilles
    in internal hatch openings (32.75)
Roller shutter doors (31.55)
    in internal openings (32.55)
Rollers (earth compaction equipment) (B6f)
Rolling (D4)
Rolling stock
    manufacturing facilities 2756
    rail transport facilities 118
Rolls H
Roof decking (27.9)
Roof doorways (37.5)
ROOF FINISHES (47)
Roof gardens 993
Roof trusses (27.9)
Roof walkways (37.8)
Roof window openings (37.4)
Roof windows (37.45)
Rooflights (37.41)
ROOFS (carcass alone, or including ceilings) (27)
    secondary elements (37)
    finishes (47)
Roof-tops 993
Room dividers and openings (32.5)
Rooms 994
Ropes J
Ropeways, aerial 113
Rot (L64)
Rot resisting materials u3
Rotproofers u3
Roughcast work P
Roughness (G4)
Routine servicing and cleaning (W2)
Row houses
    single storey housing 8114
    two storey housing 8124
Rowing facilities 546
Rubbers n5
    cement-rubber latex r4
Rubber-tyred tractors (B2f)
Rubbish disposal services (52.1)
Rubble and rubble walling F
Rugby facilities 564
Ruinous buildings (ruins) 988
Runners (temporary works) (B2d)
Running costs (Y4)
Runways 148
Rural planning areas 051
Rushes j3
Rust removing agents w1

S

Sacks (99.24)
    chute and hopper refuse disposal (52.13)
    manual refuse disposal (52.14)
    refuse disposal in general (52.19)
Sacramental facilities 68
Sacristies 68
Saddle joints Z
Saddleback roofs (27.22)
Safe deposits facilities 338
Safeguarding of site (D1)
Safes (storage fittings) (76.6)
Safety factors (U47)
Safety glass o8
Sailing facilities 546
Salaries (A2t)
Sales
    forecasts (A6f)
    income (A2s)
    representative's visits (A6gj)
Sales facilities 348
Salt marshes 092
Salt-glazed ware g3
Saltings 092
Salvage (X8)
    fire and explosions (K7)
Sanatoria 442
Sanctuaries 68
Sand p1
    cement-sand mix q2
    sandlime mix q1
Sand-blasting (D4)
Sandlime concrete f1
Sandstone e4
Sandwich panels R
Sandy soils p1
Sanitary activities (T194)
SANITARY AND HYGIENE FITTINGS
    as a whole (74)
    built-in or otherwise fixed (74)
    loose furniture and equipment (84)
Sanitary disposal fittings (74.4)
    loose furniture and equipment (84)
Sanitary facilities 94
Sanitary goods dispensers (74.7)
Sanitary suites (74.1)
Sanitation factors (U465)
Sapele i3
Sash windows (31.45)
Satellite ground stations 156
Satellite towns 053
Sauna baths 942
Sauna fittings (74.27)
Scaffolding (B2d)
Scale
    aesthetics (G1)
    relative dimensions (F47)
Scantlings H
Scenery docks 524
Scent (G8)
Scheduling (A3u)
Scheme design (A3)
SCHOOL FACILITIES 71
Schools of architecture (Ano)
Schools of building (Ano)
SCIENTIFIC FACILITIES 73
Scientific interest
    natural areas 0872
Scissor lifts (66.1)
Scrapers (B2h)
Scrapping (X8)
Screeding work P
Screening
    electrical (Q2)
    electromagnetic radiation (Q7)
SCREENING AND STORAGE FITTINGS
    as a whole (76)
    built-in or otherwise fixed (76)
    loose furniture and equipment (86)
Screening clinics 422
Screening fittings (76.7)
Screens (76.7)
Screens (grading plant) (B4d)
Screw jointed pipework I
Screwed joints Z
Screwed pile foundations (17.3)
Screwing tools (B5d)
Screws t6
Scrub W
Sculptures 984
Sculptures (fittings) (78.6)
    loose fittings (88)
Sculptures display facilities 757
Sea areas 093
    international or national in scale 02
Sea defence facilities 1863
Sea grass j3
Sea transport facilities 131
Sea walls 1863
Seabed areas 093
Sealants t4
Seals u5
Seamed joints Z
Seating (72.6)
    in external works (fixed and loose) (90.7)
    loose furniture and equipment (82)
SECONDARY ELEMENTS OF STRUCTURE (3–)
Secondary roads 1222
Secondary school facilities 713

Secretarial activities (A1s)
Secretarial offices 32
Secretarial staff (A1my)
SECTIONS AND SECTION WORK H
Secure facilities 378
Secure prisons 3762
Secure storage facilities 967
Secure storage fittings (76.6)
Securing (D6)
Securing devices t7
   or use (99.95)
SECURITY AND CONTROL
   SERVICES (68)
Security and fire protection
   services (68.1)
Security factors (U47)
Security glass o8
Security services (68.2)
Seed hairs j3
Seeds and seeding work W
Seismographic stations 737
Selection of personnel (A1mn)
Self-service shops 345
Selling (A6f)
Semi-detached buildings 982
Semi-detached housing
   single storey 8112
   two storey 8122
Semi-hydraulic binders q3
Semi-rigid joints Z
Semi-secure prisons 3762
Sensory factors (G)
Separating floors (23.8)
Separating walls (22.8)
Separation (X2)
Septic tanks (52.7)
   in external works (90.51)
Serveries (32.7)
Service lifts (66.1)
Service shops 346
Services (commercial factors) (Y9)
SERVICES (mainly electrical) (6−)
   in external works (90.6)
SERVICES (piped and ducted) (5−)
   in external works (90.5)
Services (temporary) (B2c)
   installing (D2)
SERVICES AS A WHOLE (electrical, and
   piped and ducted) (5−)/(6−)
   code (5−) alone is usually used
Services engineers (Amw)
Services heart units (5.2)
Servicing, routine (W2)
Setting agents u2
Setting out (D2)
Settlement (J12)
Settlements 05
Settling
   anti-settling agents u2
Setts and sett paving work F
Sewage and effluent treatment (52.3)
Sewage disposal facilities 174
Sewage disposal services (52.3)
Shadowing of surroundings (H2)
Shafts (solid fuel mines) 166
Shafts as part of carcass (28.8)
   lift wells (24.8)
   stair wells (24)
Shales p3
Shanty settlements 056
Shape (F1)
SHAPING (D4)
Shavings p5
Shear (J4)
Shear resistant joints Z
Sheathing section work H

Sheds 963
   in external works (90.2)
Sheep rearing and living facilities 4645
Sheet piling (17.1)
   temporary works (B2r)
Sheets and sheet work
   flexible T
   flexible proofing L
   malleable M
   rigid R
   rigid overlapping N
Shelf life (V1)
Shell roofs (27.6)
Shell structures (2.5)
Shellac v4
Shelter belt facilities 261
Sheltered housing 843
Shelters 987
   in external works (90.2)
Shelters, air raid 3757
Shelters, bus 124
Shelving (storage fittings) (76.4)
   loose furniture and equipment (86)
Sherardised coatings u1
Shingle (aggregates) p1
Shingles and shingling work N
Shipbuilding facilities 2755
Ships (water transport) 138
Shooting facilities 565
Shopping centres 342
SHOPS 34
   housing with shops 818
   road vehicle repair and parts 127
Shores (B2d)
SHORING WORK B
   used in specifications applications
Short circuits (Q11)
Short stay accommodation 448
Shovels
   earth moving tractor
     attachments (B2g)
   excavating plant generally (B2j)
Showers (cabinets, trays, curtains
   etc) (74.26)
Showers (sanitary facilities) 942
Showgrounds 583
Showrooms 758
   road vehicles 127
   shops 348
Shredders (refuse disposal) (52.18)
SHRINES 67
Shrinkage (F47)
Shrubs W
Shutter doors (31.54)
   folding (31.54)
   roller (31.55)
Shutters (screening fittings) (76.7)
Sick people (U35)
Side chapels 68
Side-effects (R6)
Side-hung doors (31.52)
   in internal openings (32.52)
Sieves (B4d)
Signal boxes 116
Signal producers (99.125)
Signalling equipment (B8r)
Signs (71.1)
   in external works (90.7)
Silica
   asbestos–silica–cement,
   asbestos–silica–lime f6
   silica based refractory materials g6
Silicon in metal alloys
   aluminium silicon h4
   silicon bronze h6

Silicon carbide based refractory
   materials g6
Silicone n5
Sills (99.83)
   for doorways (31.59)
   for window openings (31.49)
Silos 1841
   agricultural 2681
Single carriageway roads 1225
Single family housing
   see Houses
Single people (U34)
Single people's housing 847
Single storey housing 811
Single-lap tiles and tiling work N
Sink waste disposal fittings (73.2)
Sinks (99.23)
   cleaning and maintenance
     fittings (75.1)
   culinary fittings (73.2)
Sintered pulverised fuel ash p3
Sirens (99.125)
   audio communications (64.38)
   electrical services in general (69)
   fire detection and alarm (68.52)
   intruder detection and alarm (68.2)
   security and control services in
     general (68.9)
Sisal j3
Site
   beneath the structure (1−)
   external works (90)
   see also Sites
Site assembly (E4)
Site clearance (D2)
   articles found on site (A9t)
   plant (B2)
Site investigation (A3s)
Site management (A5)
Site morphology (11)
Site offices (B1b)
Site preparation (D2)
   articles found on site (A9t)
   plant (B2)
Site protection (D1)
   plant (B1a)
Site roads (B2c)
Site walls (external works) (90.3)
SITES (construction projects as a
   whole) (−−)
   see also Site
Siting factors (F1)
Sixth form colleges 714
Size (F4)
Size (seals) u5
Skating facilities 567
Ski jumps and slopes 567
Ski lifts 113
Ski school facilities 567
Skid pads 128
Skills (Am)
Skim coating work P
Skimmers (B2j)
Skinning
   anti-skinning agents u2
Skip dumpers (B3e)
Skips (B3x)
Skirtings (43)

**Consult the schedules to check
the context of items in the index.
Italic entries are explained in
the introduction.**

Sky-Sta

Sky factor (N7)
Skylights (37.45)
Skyscrapers 981
Slab and column frames (28.3)
Slab and panel structures (2.3)
Slab floors (suspended) (23.2)
    floor beds (13)
Slab plus beam floors (23.4)
Slab urinals (74.45)
Slabs
    rigid tile work S
    thin slabs (rigid sheet work) R
Slabs in building frames (28.9)
Slag, blast furnace
    air cooled p2
    foamed p3
Slag cement q2
Slag heaps 176
Slate e5
    aggregate p3
Slates and slating work N
Slatings H
Sleep facilities 921
Sleep fittings (72.1)
    loose furniture and equipment (82)
Slenderness (F47)
Sliding door gear (31.54)
    for wardrobe doors (76.2)
Sliding doors (31.54)
    in internal openings (32.54)
    in wardrobes (76.2)
Sliding joints Z
Sliding partitions (32.54)
Sliding poles (24.6)
Slings (B3x)
Slipper baths 942
Slipperiness (J5)
Slipways (roads) 128
Slipways (water transport) 138
Slopes (11.1)
Sloping roofs (27.21)
Sluice rooms 951
Slums 0833
Slurries Y
Smell (G8)
Smog (H2)
Smoke (K5)
Smoke chambers 372
Smoke detectors (68.52)
Smoothing (D4)
Smoothness (G4)
Snack bars 515
Snooker facilities 562
Snow (H122)
Soakers (99.85)
    roof finishes (47)
Soap, liquid (piped and ducted
    services) (53.8)
Soap dispensers (units and
    systems) (74.7)
Soaps w7
Social activities (T151)/(T153)
Social economics (U2)
Social psychology (U2)
SOCIAL RECREATION FACILITIES 53
Social refreshment facilities 51
Societies (U1)
    construction industry (Ak)
Societies, learned
    facilities 727
Society (U2)
Sociological factors (U2)
    characteristics of areas 097

Socket outlets (99.365)
    communications services (64.9)
    electrical power subcircuits (62)
    electrical services in general (69)
Soft furnishings (78.3)
    curtains and drapes (76.7)
Soft surfaces
    planted and unplanted beds (13.4)
    in external works (90.42)
Softened water supply (53.6)
Softeners, water u2
Softwood i2
Soil and waste water disposal (52.3)
Soil conditioners w6
Soil erosion protection 1863
Soil heating (external works) (90.52)
Soil mechanics (J12)
Soil stabilization equipment (B6h)
Soil testing equipment (B2q)
Soils p1
Solar control glass o6
Solar energy supply facilities 163
Solar heat gain (M5)
Solar heating (56.15)
Solar radiation (Q7)
Solar radiation as power source (5.15)
Solder jointed pipework I
Soldered joints Z
Soldering materials t2
Soldiers (grounds, for fixing) (B2d)
Solid bricks F
Solid fuel as power source (5.11)
Solid fuel extraction supply facilities 166
Solid fuel heating (56.11)
Solid fuels w3
Solids (formless products) Y
Solids (interactive factors) (L4)
    mechanics (J1)
Solids (state of substances) z1
Solubility (L37)
Solutions (L3)
Solvency (L37)
Solvents u2
    organic solvent preservatives u3
Soot (K5)
Sorting offices 157
SOUND (PROPERTIES AND
    PROCESSES) (P)
Sound attenuators (99.535)
    air conditioning and ventilating
    services (57.9)
    sound control services (68.8)
Sound control services (68.8)
Sound insulation (P2)
Sources of information (Agm)
SPACE COOLING SERVICES (55)
Space frame roofs (27.6)
Space frames as building frames (28.6)
SPACE HEATING SERVICES (56)
Space resources (U6)
Spaces 997
    between buildings 998
Span (F47)
Spare parts and accessories shops 127
Spastic people
    see Handicapped people
Spatial relationships (F1)
SPECIAL ACTIVITY FITTINGS
    as a whole (77)
    built-in or otherwise fixed (77)
    loose furniture and equipment (87)
Special cross sections H
Special school facilities 717
Special shape bricks F
Special surface coating work P
Specific heat (M4)

Specifications (A3u)
    British Standard specifications (Ajr)
Specifying (A3u)
Speculative housing 818
Speed (R1)
Spigot and socket joints Z
Spiral stairs (24.4)
Spires (27.5)
Splitting (J4)
Sporting activities (T154)/(T156)
SPORTS FACILITIES 56
    aquatic sports 54
Sports fittings (77)
    in external works (fixed and
    loose) (90.7)
    loose furniture and equipment (87)
Spot lighting (63.2)
Sprayed asbestos cement q9
Sprayed protective coatings u1
Spray-painting equipment (B5h)
Spread of fire
    flame spread (K45)
    prevention (K22)
SPREADABILITY (V6)
Spreaders (B6c)
Sprinklers (fire fighting services) (68.54)
Spruce i2
Squares (urban areas) 058
Squash facilities 562
Squeezed joints Z
Stabilised ground (11.3)
Stabilisers u2
Stability (X5)
    mechanics (J7)
Stables 4643
Stack bond blockwork and brickwork F
Stadia 564
Staff
    see Personnel
Staff costs (A2t)
Staff location services (communications)
    audio (64.35)
    visual and audio-visual (64.13)
Staff rooms 922
Staff training (A1n)
    costs (A2t)
Stages (entertainment facilities) 528
    drama 524
Stages (secondary suspended
    floors) (33.1)
    platforms (71.3)
Staining, pattern (L44)
Stainless steel h3
Stains v2
STAIR FINISHES (44)
Stair nosings (44)
Stair treads
    carcass (24.9)
    finishes (44)
Stair wells (24)
STAIRS AND RAMPS (carcass alone, or
    including soffites) (24)
    secondary elements (34)
    finishes (44)
Stairways 9142
Stakes, horticultural (B8w)
Stall urinals (74.45)
Stalls, trading 348
Stanchions (99.751)
    in building frames (28.9)
Standard factories 282
STANDARDIZATION (Ap)
Standards (Ajv)
    British Standard specifications (Ajr)
Standby generators (61.7)
Stands, sports 568

State school facilities 718
States of substances z1
   properties (E1)
Static (electrical charge)
   anti-static agents u9
Static electricity factors (Q12)
Static tower cranes (B3t)
Statics (J)
Stationery (A1s)
   photocopying etc. (A1t)
Stations, railway 114
Statutes (Ajk)
   Statutory Instruments, if kept
     separately (Ajm)
Statutory requirements (Ajk)
   designing and physical planning (A3j)
   financing (A2j)
   inspection and quality control (A7j)
   project (or office) administration (A1j)
Steam boilers (B8h)
Steam supply and condensate (54.2)
Steel h2
Steel alloys h3
Steeples (27.5)
Stem fibres j3
Step irons (24.6)
Steps (in external works) (90.41)
Steps, sets of (71.3)
Stiffening (X7)
Stock control (A6s)
Stock exchanges 337
Stock rooms 961
Stock-broking facilities 337
STONE e
   artificial stone f3
   crushed stone p1
   stone-bitumen mixes s5
Stone block work F
Stoneware g2
   glazed stoneware g3
Stop ways (air transport) 148
Stopping (fillers) v1
Stopping, fire (K22)
Stops
   bus 124
   railway 114
Storage activities (T196)
STORAGE AND SCREENING FITTINGS
   as a whole (76)
   built-in or otherwise fixed (76)
   loose furniture and equipment (86)
Storage facilities, bulk goods 184
   agricultural 268
   gas 165
   oil 164
   refuse 1751
   sewage 1741
   water 1711
Storage facilities, common 96
Storage facilities, culinary 935
Storage facilities, industrial 284
Storage facilities, transport
   aircraft 147
   boats 137
   common facilities 149
   rail vehicles 117
   road vehicles 127
Storage fittings, culinary (73.5)
   loose furniture and equipment (83)
Storage heaters (electrical central
   heating) (56.6)
Storage life (V1)
STORAGE PLANT (B4)
Storage walls (76.1)
Stores (shops) 34
Stores, costs of (A2t)

Storeys 993
Storing
   ease of storing (V1)
Storms (H122)
Stoves (heating appliances) (56.8)
Straddle trucks (B3h)
Straight stairs (24.1)
Strain energy (R2)
Straw j3
   vegetative products w6
Strawboard j3
Streams 093
Street furniture 128
Street parking 125
Streets 12
   built-up areas 058
Strength (J3)
   impact strength (J4)
Stress (health) (U467)
Stressed skin structures (2.4)
Stresses (J4)
Stretcher bond blockwork and
   brickwork F
Stretcher/header bond blockwork and
   brickwork F
Strings J
Strip foundations (16.4)
Strips
   flexible T
   flexible proofing L
   malleable M
   sections H
Strong rooms 967
Struck joints Z
Structural engineers (Amw)
Structural mechanics (J11)
Structural members (99.7)
Structural section work H
Structure, internal (E1)
STRUCTURE ELEMENTS AS A
   WHOLE (2−)/(4−)
   code (2−) alone is usually used
STRUCTURE FINISHES (4−)
Structure plans 038
STRUCTURE PRIMARY ELEMENTS (2−)
STRUCTURE SECONDARY
   ELEMENTS (3−)
Structures (2.1)/(2.8)
   in external works (90.2)
Struts (99.751)
Struts (temporary works) (B2d)
Strutting (D2)
Stud partitions (22.6)
Students (technical staff) (A1my)
Students unions 534
Studios 32
   entertainments production
     studios 528
Studs t6
Study facilities 768
Study tours (Anq)
Study-bedrooms 921
Style (G1)
Subcircuits (99.362)
   power (62)
   lighting (63)
Submersible pumps (B2t)
Subnormal people
   see Handicapped people
Sub-regional planning areas 033/038
Subsidence (J12)
   physiographic factors (H16)
SUBSTANCES z
Substations 162
Substitution (X4)
SUBSTRUCTURE (1−)

Subtraction (X2)
Sub-tropic factors (H11)
Suburban railways 111
Suburbs 058
Subways 181
Suction piped supply of solids (54.8)
Suction pipes I
Suitability (T3)
Sulphate resisting cement q2
Sunlight (N11)
Sunshine (H122)
Supermarkets 344
Superstructure (primary elements) (2−)
Supersulphated cement q2
Supply (Y8)
Supports (99.76)
Surface areas 092
   international or national in scale 02
Surface coefficient (thermal
   transmission) (M31)
Surface hardeners u5
Surface railways 111
Surface resistance (thermal
   transmission) (M31)
Surface temperature (M7)
Surface treatments, ground (in external
   works) (90.4)
   floor beds (13)
Surface vibrators (B6g)
Surface water drainage (52.5)
Surgeries 423
   special surgeries 424
Surgical facilities in hospitals 4172
Surroundings (H)
Surrounds (99.87)
Surveillance security services (68.2)
Surveying (A3s)
Surveyors (Amw)
   central and local government
     departments (Aks)
Suspect work (A7r)
SUSPENDED CEILINGS (35)
   finishes (45)
Suspended floors (carcass alone or
   including ceilings) (23)
   suspended ground floors (13.1)
   secondary elements (33.1)
   finishes (43)
Suspended roofs (27.6)
Suspended storage fittings (76.5)
   loose furniture and equipment (86)
Suspending agents u2
Suspension bridges 1828
Swimming facilities 541/544
Swimming pools (in external
   works) (90.8)
Swing bridges 1825
Switches (99.963)
   electrical services in general (69)
   lighting subcircuits (63.9)
   power subcircuits (62)
Switchgear (99.416)
   communications services (64.9)
   electrical supply services (61.9)
Switching facilities
   (telecommunications) 156
Synagogues 65

Consult the schedules to check the context of items in the index. Italic entries are explained in the introduction.

Synchronous clock services (64.5)
Synthetic adhesives t3
Synthetic fibres n6
Synthetic resins n5/n6
    compounds and mortars r4
    emulsion paint v6
    oil paints v5
Synthetic rubbers n5
Systems, building (– –)
Systems theory (Af)
    project (or office) administration (A1f)

T

T joints Z
T sections H
Table tennis facilities 562
Table ware (83)
Tables (72.6)
    dining tables (73.7)
    loose furniture and equipment (82)
Tactical plans 038
Tactile factors (G8)
    texture (G4)
Tailors 346
Tall buildings 981
Tamping (D6)
Tanking (99.532)
    in floor beds (13.9)
    in basement retaining walls (16.2)
    in substructure generally (19)
Tanking coating work P
Tanks (99.23)
    drainage services (52.9)
    liquids supply services (53.9)
    piped and ducted services in general (59)
Tar s1
Tanks (storage plant) (B4k)
Tanks facilities 1841
    agriculture 2681
    water 1711
Tap rooms 517
Taper pipes I
Tapes H
Taps (99.411)
    liquids supply services (53.9)
    gases supply services (54.9)
    piped and ducted services in general (59)
Tar s1
Tar macadam s5
Tar oil preservatives u3
Taste (G8)
Taverns 517
Taxation (A2j)
Taxation facilities 315
Taxi ways (air transport) 148
Teaching activities (T178)
Teaching facilities 729
    further education 728
    schools 718
Teaching hospitals 411
Teak i3
Team ball games facilities 564
Team teaching facilities 7294
Tearing (J4)
Technical colleges 722
Technical literature preparation (Ah)
Technical school facilities 713
Technicians (Amv)
    project (or office) administration (A1my)
Technological universities 721
Teenagers (U31)

Telecommunications facilities 151/156
Telegraph facilities 1542
Telegraph services (elements) (64.41)
Telephone booths (64.32)
    in external works (90.64)
Telephone exchanges (services elements) (64.32)
Telephone facilities 1541
Telephone services (elements) (64.32)
Telephones
    project (or office) administration (A1s)
Teleprinter services (elements) (64.42)
Telescopic shaft lifts (66.1)
Television facilities 153
Television services (elements) (64.11)
Telex
    project (or office) administration (A1s)
Telex facilities 1542
Telex services (elements) (64.42)
Tellers' boxers 338
Temperate areas 095
Temperate factors (H11)
Temperature (M7)
    climate (H121)
TEMPLES 65
Temporary buildings 983
TEMPORARY (NON-PROTECTIVE) WORKS (B2)
Temporary residential facilites 87
Temporary services (B2c)
    installing (D2)
Tenancy (U2)
Tenanted areas 096
Tenants (U36)
Tendering (A4t)
Tennis facilities 562
Tennis courts (in external works) (90.8)
Tenpin bowling alleys 563
Tension (J4)
Tension structures (2.6)
Terminals
    air transport 144
    railways 114
    water transport 134
Terminology (Agh)
    documentation (Ahi)
Terra cotta g2
Terraced housing
    single storey 8114
    two storey 8124
Terraced roofs (27.1)
Terraced slopes (11.1)
Terraces 982
Terrazzo q5
    precast f3
Tesserae S
TESTING (Aq)
    quality control (A7q)
Testing equipment (B8v)
    soil (B2q)
Textile manufacturing facilities 2761
Texture (G4)
Textured glass o2
Texturing (D4)
Thatching work K
Theatres 524
Thermal insulation (elements) (99.534)
    for cavity walls (21.1)
    for floors (23.9)
    for roofs (27.9)
    for structure generally (29)
    for suspended ceilings (35.9)
    for walls generally (21.9)

Thermal processes and properties (M)
Thermal value (K42)
Thermic boring equipment (B5e)
Thermoplastics n6
Thermosets n6
Thermostats (99.413)
    air conditioning and ventilation services (57.9)
    space cooling services in general (55.9)
    space heating services in general (56.9)
THICK COATING WORK P
Thickness (F47)
Thin slabs R
Thinners u2
Three coat work P
Three storey housing 814
Thresholds (for doorways) (31.59)
Thrust resistant wall storage structures 1842
    agricultural 2682
Tides (H16)
Tied retaining walls (16.2)
Ties (99.752)
    for cavity walls (21.1)
Ties, horticultural (B8w)
Tiles and tiling work
    flexible T
    flexible proofing L
    malleable M
    rigid S
    rigid overlapping N
Tilt dozers (B2g)
Timber i
Timber industries facilities 2772
Timbering (D2)
TIME (Z)
    effect of delay (A5r)
Tippers (B3d)
Tips, refuse 1753
Tobacco manufacturing facilities 273
Toboggan runs 567
Toilets 941
Tolerance (F6)
Tolerance (behaviour) (U2)
Tombs 67
Tongued and grooved joints Z
TOOLS (B)
Top coating work V
Topographical factors (H15)
Topping coating work P
Torsion (J4)
Tort (Ajk)
Total thermal resistance (M31)
Toughened glass o8
Toughening (X7)
Towel cabinets and rails (74.3)
Tower cranes (B3t)
Tower derricks (B3t)
Towers 183
    air traffic control 146
    cooling towers 161
    fire practice 372
TOWN AND COUNTRY PLANNING 05/08
Town centres 058
Town halls 314
Town houses, historical 86
Towns 053
    expanding 055
    new, utopian 054

Townscape 058
Towpaths 123
Toxicity (U466)
Track and field events facilities 564
Tracks
  cycle 1232
  sports 568
Tracks, railway 118
  transport plant (B3k)
Tractor attachments (B2g)
Tractors (B3g)
  earth-moving plant (B2f)
Trade associations (Akp)
  publications (Ajv)
Trade literature preparation (Ah)
Trade unions (Akp)
Trading (accounting procedures) (A2r)
Trading activities (T134)
TRADING FACILITIES 34
Traffic controls 126
Traffic factors (H2)
Traffic lanes 128
Trailers (B3d)
Training (An)
  staff training (A1n)
  staff training costs (A2t)
Training facilities
  physical training 562
  sports 568
Transepts 68
Transfer (X3)
Transformation (X4)
Transformers (99.136)
  electrical supply services (61.9)
Transformers (construction plant) (B8e)
Transit sheds (water transport) 134
Translucency (G6)
Translucent glass o2
Transmission
  energy (R5)
  light (N3)
  sound (P3)
  thermal (M31)
Transmission facilities
  electricity 162
  telecommunications 156
Transmissivity (M31)
Transmittance (M31)
Transmitters (99.32)
  communications services (64.9)
Transmitting stations 156
Transparency (G6)
Transparent glass o1
Transport and communication
  facilities 11/15
TRANSPORT FACILITIES 11/14
TRANSPORT PLANT (B3)
TRANSPORT SERVICES (elements) (66)
Transporter bridges 1825
Trap doors and doorways
  floor secondary elements (33.5)
  roof secondary elements (37.5)
TRANSPORTING (D3)
Traps (99.142)
  drainage services (52.9)
Travel facilities 11/14
Travelling cradles (66.3)
Treasuries, cathedral 62
TREATMENT (CONSTRUCTION
  OPERATIONS) (D5)
Treatment facilities
  refuse 1752
  sewage 1742
  water supply 1713
Tree grilles (90.41)
Trees W

Trenchers (B2n)
Trenches (11.9)
Trenching (D2)
Tribes (U1)
Trims (99.98)
  finishes to structure (49)
  *ground surface treatments (external
    works) (90.41)*
Trolleys, piano (B3i)
Tropic factors (H11)
Trough lighting fittings (63.9)
Troughed sheets and sheet work N
Troughs (washing fittings) (74.23)
Truck mounted cranes (B3q)
Truck mounted tower cranes (B3t)
Trucks (B3d)
  dump trucks (B3e)
  platform, fork-lift, straddle
    trucks (B3h)
Trunk roads 1221
Trunking (99.361)
  as part of suspended ceilings (35.9)
  electrical services in general (69)
  lighting services (63.9)
Trunking work I
Trusses (99.7)
  for floors (23.9)
  for roofs (27.9)
  for floors and roofs (29)
Tubers W
Tubes (hollow sections) H
Tung oil v3
Tunnels 181
  rail 118
  road 128
Tunnels (elements) (11.5)
Turfing work W
Turkish baths 942
Turntables (conveyors) (66.5)
Twilight areas 0833
Twisting (D4)
Two brick brickwork F
Two coat work P
Two part synthetic resin mortars r4
Two storey housing 812
Typography (Ahm)
Tyrolean coating work P

U

U value (M31)
UN facilities 311
Ultra-violet absorbers u9
Ultra-violet radiation (Q7)
Uncoursed blockwork F
Uncovered spaces 996
Undercoating work V
Underground areas 094
  international or national in scale 02
Underground buildings 988
Underground elements (11.5)
Underground railways 112
Underlays (99.85)
Underpasses 181
Underpinning (B2d)
Underpinning work B
  used in specifications applications
Underwater areas 093
  international or national in scale 02
Underwriting facilities 335
Unglazed fired clay ware g2
Unions (students' facilities) 534
Unions (workers trades
  organisations) (Akp)
Unit air conditioning (57.3)

Unit ventilation (57.7)
United Nations facilities 311
Universities 721
Unloading (D3)
Unmarried mothers' welfare
  facilities 445
Unmarried people (U34)
Unorganized groups (U1)
Unreinforced casting work E
Unusual plants W
Unwrot timber i1
Up and over doors (31.56)
Updating procedure (network
  analysis) (A5t)
Upholsterers 346
Upholstery (78.3)
  loose furniture and equipment (88)
Urban planning areas 052/058
Urea resins n6
Urethane n5
Urinal facilities 944
Urinals (74.45)
Use
  application (T)
  costs in use (Y4)
  occupancy of facilities (H7)
Usefulness (T4)
Users (U)
  user requirement studies (A3r)
UTILITIES, PUBLIC 1
Utility (application factors) (T)
Utility rooms 955
Utopian towns 054

V

V joints Z
VTOL facilities 142
Vacuum cleaning fittings (75.8)
  loose furniture and equipment (85)
Vacuum refuse disposal (52.13)
Vacuum supply services (54.5)
Valleys 092
Valuations (A4x)
  work in progress (A2r)
Values (colour) (G5)
Values (economics) (Y6)
Valves (99.411)
  gases supply services (54.9)
  space heating services (56.9)
  *piped and ducted services in
    general (59)*
Vandalism (L61)
Vandals (U37)
Vanity basins (74.23)
Vaporising liquid fire fighting
  services (68.53)
  automatically operated (68.54)
Vapour barriers (99.532)
  external walls (21.9)
  roofs (27.9)
  structure generally (29)
Vapour proof flexible sheets L
Vapour supply services (54.2)
Vapours (L2)
  explosive limits (K6)
Variations (A4w)

**Consult the schedules to check
the context of items in the index.
Italic entries are explained in
the introduction.**

Var–Whe

Varnishes v4
Vaulted roofs (27.6)
Vaults (secure storage facilities) 967
Vaults, funeral 67
VEGETABLE MATERIALS j
 glues t3
Vegetables W
Vegetative products w6
Vehicle bridges 1822
Vehicle manufacturing facilities 2756
Vehicle transport facilities 11/12
 common facilities 149
Vehicular hardstandings (90.41)
Vending machines (catering
 fittings) (73.8)
Veneers and veneering work T
 wood veneers i5
Ventilation (L2)
VENTILATION AND AIR CONDITIONING
 SERVICES (57)
Ventilation bricks and brickwork F
Ventilation plant (B8b)
Ventilation services (57.6)/(57.7)
 central ventilation (powered
 supply/extract) (57.6)
 unit ventilation (direct, locally
 acting) (57.7)
 *code (57.6) is used as the general
 position*
Ventilation systems
 drainage services in general (52.9)
Venting
 explosions (K6)
 fire (K23)
Verandahs 996
Vermiculite m1
 exfoliated vermiculite p3
Vermin (L62)
Vertical boom trenchers (B2n)
Vertical circulation elements (24)
Vertical dividing elements (21)
Vertical lift bridges 1825
Vertical take-off and landing facilities 142
Vestibules 913
Vestries 68
Veterinary hospitals 461
Viaducts 1826
Vibrating rollers (B6f)
Vibration (J6)
Vibrators (concrete compaction
 equipment) (B6g)
Village halls 532
Villages 051
Violence (U2)
Viruses (U466)
Viscose rayon n6
Viscosity (L5)
 mechanics (J5)
Visibility (G1)
Visitors (U37)
Visual communications services (64.1)
Visual factors (G)
Vitreous china g3
Vitrified clay g2
Volcanic areas 095
Volcanic eruptions (H16)
Voltage (Q5)
Volume (F45)

W

WC facilities 944
WC pans, seats, covers, cisterns (74.44)
WC suites (74.44)
Wages (A2t)
Wagons (transport plant) (B3i)
Wagons, railway 118
 transport plant (B3k)
Waiting facilities 913
Walkways
 roof secondary elements (37.8)
 suspended ceilings (35.8)
WALL FINISHES, EXTERNAL (41)
WALL FINISHES, INTERNAL (42)
WALL FINISHES AS A WHOLE (41)/(42)
 *Code (41) alone can be used for library
 purposes; (4–) is used for project
 documents*
Wall plates (21.9)
Wall ties (21.1)
Wallcoverings, internal (42)
Walling
 blocks, bricks and stone F
 rigid sheets R
Wallpapers T
Walls
 city 3756
 sea 1863
WALLS (elements) (21)
 external walls (21)
 internal walls (22)
 retaining walls (16.2)
 rising walls (footings) (16.4)
 site walls (external works) (90.3)
Walls, non-thrust resistant
 agricultural storage structures 2683
 storage structures generally 1843
Walls, thrust resistant
 agricultural storage structures 2682
 storage structures generally 1842
Walnut i3
Warden's houses 841
Wardrobes (76.2)
 loose furniture and equipment (86)
Wards 418
Warehouses 284
Warm air central heating (56.5)
Warmth to touch (G8)
Warning notices (B1c)
Wash basins (74.23)
Wash houses 952
Washbays 127
Washing facilities 951
Washing fittings
 cleaning and maintenance
 fittings (75.1)
 culinary fittings (73.2)
 sanitary and hygiene fittings (74.2)
Washing-up facilities 932
Washrooms 942
WASTE DISPOSAL AND DRAINAGE
 SERVICES (52)
Waste disposal units
 culinary fittings (73.2)
 sanitary and hygiene fittings (74.48)
Waste land 0834
Waste matter as heat source (56.16)
Waste matter as power source (5.16)
Waste of resources (T6)
Waste packaging units (52.18)
Watching galleries 68
Water (elements of construction projects)
 in external works (90.43)
 in floor beds (13.6)

Water (materials) w4
Water areas 093
Water bars (99.532)
 in doorways (31.59)
Water closets, *see* WC
Water coolers (catering fittings) (73.8)
Water factors (L3)
 'clean' water (H17)
Water features 998
Water flow control facilities 187
Water meters (53.9)
Water paints v8
Water plants W
Water reducing agents u2
Water repellants u6
Water retaining agents u2
Water retaining elements (16.3)
Water retention facilities 187
Water ski-ing facilities 548
Water softeners u2
Water supply facilities 171
 common facilities 972
WATER TRANSPORT FACILITIES 13
Water treated for special purposes (53.6)
Water vapour (L2)
Water-borne preservatives u3
Water-borne refuse disposal (52.3)
Watermills 163
Waterproofing (requirements and
 properties) (L34)
Waterproofing coating work
 film V
 thick P
Waterside plants W
Waxes Y
Ways of life (U2)
Weakening (X7)
Wealth (U2)
Wealthy people (U36)
Wear (X7)
 mechanics (J4)
Weather (H12)
Weather proofing (H123)
Weatherbars
 for doorways (31.59)
Weathered joints Z
Weathering (H14)
Wedge shape blocks F
Weedkillers w9
Weight (F5)
Weight lifting facilities 562
Weirs 187
Weldability (V6)
Welded joints Z
Welding (D4)
 equipment (B5e)
Welding materials t1
Welfare activities (T144)/(T146)
Welfare facilities 44
 animals 46
Welfare of personnel (A1mr)
Wells
 gas 165
 oil 164
 water 1711
Welted joints Z
Wet weather (H121)
Wetting agents u2
Wharves 138
Wheel bogeys (B3i)
Wheel sports facilities 564
Wheel type boom trenchers (B2n)
Wheelbarrows (B3i)
Wheeled cranes (B3q)

Wheeled tractors (B3g)
    earth-moving plant (B2f)
White cement q2
White metal h8
Whitewash v8
Whitewood i2
Wholesaling facilities 341
Width (F47)
Wild animals (requirements) (U51)
Winches
    earth-moving tractor
      attachments (B2g)
      hoisting plant (B3x)
      pile driving equipment (B2r)
      transport plant tractor fixtures (B3g)
Wind barriers (99.531)
Wind effects around tall buildings (H2)
Wind loads (J4)
Windmills 163
Window and door furniture (31.9)
Window fittings (31.49)
Window frames (31.49)
Window openings (31.4)
    in internal walls (32.4)
    in roofs (37.4)
Window sills (31.49)
Windowless buildings 988
Windows (31.4)
    internal windows (32.4)
Winds (H122)
Wine cellars 935
Wings (building divisions) 994
Winter sports facilities 567
Wire cut bricks F
Wire netting backing work J
Wired glass o4
WIRES AND WIRE WORK J
Witness box 317
Women (U33)
WOOD i
    fibres j1
    particles j7
Wood wool–cement j8
Woods (forests) 092
    forestry facilities 261
Wool j6
WORK AND REST FITTINGS
    as a whole (72)
    built-in or otherwise fixed (72)
    loose furniture and equipment (82)
Work benches (72.3)
    loose furniture and equipment (82)
Work facilities 923
Work fittings, culinary (73.1)
    loose furniture and equipment (83)
Work fittings, general purpose (72.3)
    loose furniture and equipment (82)
Work study (A5p)
Work tops, cupboards and drawers for
    kitchens (73.1)
Workability (V)
Workability aids u2
Workers (U37)
Working activities (T192)
WORKING FACTORS (V)
Works of art (78.6)
    loose furniture and equipment (88)
Workshop equipment (construction
    plant) (B5j)
Workshop fittings (77)
    loose furniture and equipment (87)
Workshops 282
Worshipping activities (T16)
Wrestling facilities 562
Wrong use (T7)
Wrot hardwood i3
Wrot softwood i2
Wrought iron h1

X

X-ray absorbing glass o6
X-ray facilities in hospitals 4171
X-ray rejecting glass o6
X-rays (Q7)

Y

YMCA hostels 856
Yards 2688
Yardsticks (costs) (A4j)
Yarns J
Yield point (J7)
Youth centres 534
Youth hostels 856
Youths (U31)

Z

Z sections H
Zinc h7
    aluminium zinc h4
Zoos 751

Consult the schedules to check the context of items in the index. Italic entries are explained in the introduction.

Applications

1 Project information

1.1 Introduction

Everybody associated with CI/SfB co-ordinated projects needs to have a brief explanation of the way the system is used for the arrangement of the project documents. This may be bound in at the beginning of the bill of quantities or specification, shown on the first drawing issued, or given on a separate sheet. The specimen explanation shown in Figure 1 may be adapted as necessary, and should be reinforced by a meeting with the people receiving project documents. It is included here because it also serves to outline the system to producers of co-ordinated sets of documents.

It is very important that the first use of the system for project information should be on a reasonably straightforward (though not a very small) project, and should be kept simple. For example, any temptation to produce a drawing automatically for every element heading **must** be resisted.

Many location drawings will get a broad Table 1 code eg: (5–). Advice on setting up a pilot project, a description of the main categories used for the arrangement of project information and detailed information on alternative methods was given in the 1971 CI/SfB Project Manual. That manual also set out the advantages of coordinated project documentation. The headings which follow outline the key applications from which the greatest benefit may be obtained.

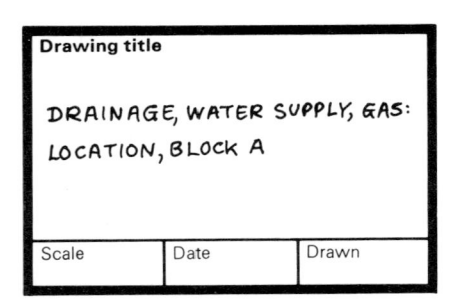

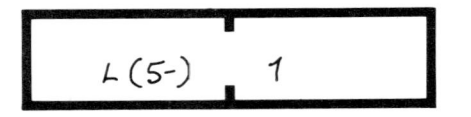

Figure 1
Specimen introduction to a set of CI/SfB arranged project documents.

```
┌─────────────────────────────┐
│ Drawing title               │
│                             │
│ DRAINAGE, WATER SUPPLY, GAS:│
│ LOCATION, BLOCK A           │
│                             │
├──────┬──────┬───────────────┤
│Scale │Date  │Drawn          │
└──────┴──────┴───────────────┘
```

```
┌──────────────┬──────┐
│   L(5-)      │  1   │
└──────────────┴──────┘
```

1a
Drawings title box.

1b
CI/SfB Table 1, as presented in BRE Digest 172.

DRAWINGS

The drawings for the project are divided into three series: Location drawings (L series) showing the position of work; Assembly drawings (A series) showing fixing information; and Component drawings (C series). The filing number is in the bottom right hand corner of each drawing, Figure 1a.

L series drawings
The first L series drawings in the set are general plans, elevations and sections, numbered L(--)1 (or L1) etc. Others are more specific and have filing numbers which include single numbers in brackets eg: L(5-), limited to any or all of the elements listed in column (5-), in Table 1, Figure 1b. Drawings with two numbers in brackets eg: L(52) are even more specific. Schedules which show the location of work are included in the L series. All L series drawings so far produced for the project are listed at the end of this note, and should be filed in the order given.

A and C series drawings
Each series is numbered by Table 1, Figure 1b.
A series drawings eg: A(22) show how components are to be brought together and fixed. C series drawings eg: C(22) show how individual components are to be manufactured. All A and C series drawings so far produced for the project are listed at the end of this note, and should be filed in the order given.

(- -) Site, project								
Substructure	**Superstructure**			**Services**		**Fittings**		**Site**
(1–) Ground, substructure	(2–) Primary elements	(3–) Secondary elements	(4–) Finishes	(5–) Mainly piped	(6–) Mainly electrical	(7–) Fixed	(8–) Loose	(9–) External elements
(10)	(20)	(30)	(40)	(50)	(60)	(70)	(80)	(90) External works
(11) Ground	(21) External walls	(31) External openings	(41) External	(51)	(61) Electrical supply	(71) Circulation	(81) Circulation	(91)
(12)	(22) Internal walls	(32) Internal openings	(42) Internal	(52) Drainage, waste	(62) Power	(72) Rest, work	(82) Rest, work	(92)
(13) Floorbeds	(23) Floors	(33) Floor openings	(43) Floor	(53) Liquid supply	(63) Lighting	(73) Culinary	(83) Culinary	(93)
(14)	(24) Stairs, ramps	(34) Balustrades	(44) Stair	(54) Gases supply	(64) Communications	(74) Sanitary	(84) Sanitary	(94)
(15)	(25)	(35) Suspended ceilings	(45) Ceiling	(55) Space cooling	(65)	(75) Cleaning	(85) Cleaning	(95)
(16) Foundations	(26)	(36)	(46)	(56) Space heating	(66) Transport	(76) Storage, screening	(86) Storage, screening	(96)
(17) Piles	(27) Roofs	(37) Roof openings	(47) Roof	(57) Ventilation	(67)	(77) Special activity	(87) Special activity	(97)
(18)	(28) Frames	(38)	(48)	(58)	(68) Security, control	(78)	(88)	(98)

Complete details:
plan, section, elevation

✓ See A(16)5   ✓ See C(76)2

Sections

See A(22)8

assembly detail

See L(--)2

view

Plans

See A(2-)12

assembly detail

See L(--)4

detailed layout

Elevations

See L(--)5

Specifications and bills

✓ blockwork: spec F11/A5

specification item

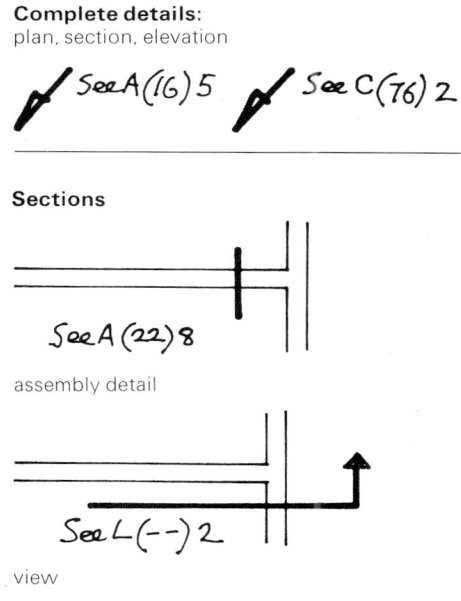

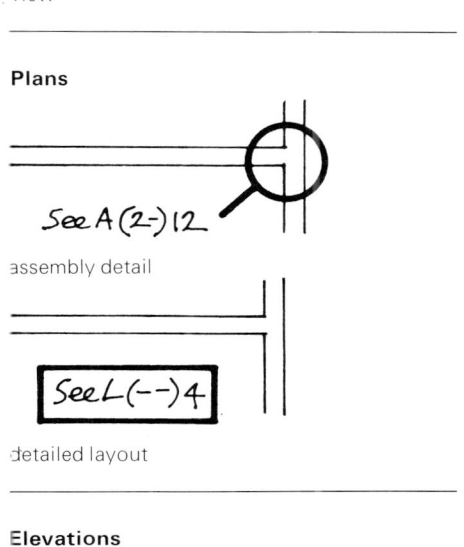

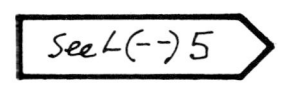

✓ blockwork: bill (21)
     spec F11/A5

elemental bill and specification item

1c
Cross reference symbols.

Finding information on drawings
To find information, look first at the relevant small scale plans, elevations and sections in the L series. These will usually include direct cross references to drawing(s) which give more detailed information. Symbols used for cross reference are given in Figure 1c. Where no cross reference is given look for L, A or C series drawings which are numbered according to the bracketed numbers given under the relevant headings in Table 1, eg: for information on stairs, look at all drawings numbered (24) and (2-); and at those numbered L1, L2 etc. If the information wanted is not on drawings with those numbers, it is probable that it has not been given at all.

SPECIFICATION AND QUANTITIES

(Producer to add information here on method of arrangement, see Project Information 1.3)

CROSS REFERENCE SYMBOLS

Figure 1c shows cross reference symbols used on the project.

CI/SfB
Table 1 (Figure 1b) is based on CI/SfB, the UK version of the international SfB system. CI/SfB is also used in the arrangement of product information and for other purposes in the construction industry. Any questions on the arrangement of the project documents will be dealt with by Mr ...... at this office, telephone ......

LIST OF DRAWINGS FOR THIS PROJECT

(Producer to add here lists of L, A and C drawings in filing order)

Application 1: Project information

1.2 Design

1.21 Briefing and outline design

CI/SfB can be used as a check list for collecting and arranging briefing information to prepare an outline technical specification. This may be used in turn for the preparation of initial cost plans and for building regulations approval, and leads on to the preparation of detailed design drawings, working drawings and specifications. Figure 2 suggests CI/SfB tables useful for these purposes.

Figure 2

Briefing information (arrange by Tables 0 and 1)

Outline technical specification (arrange by Table 1)

| Add to outline technical specification information required for Building Regulations approval (arranged by Table 1) | Detailed design drawings, Working drawings (arrange by Table 1) and Specification (arrange by Tables 2 and 1) | Add to outline technical specification information required for Cost Plan (arrange by Table 1) |

A simple specimen check list of spaces for organising briefing information, based on Table 0 and applicable to a primary school project, is shown in Figure 3. Other check lists can be produced for particular building types as required, using Table 0.

Figure 3
Specimen check list

Briefing file: Cover

712 appears on the cover of the briefing file and fixes its eventual filing position in the office library or library of past projects. The number can be followed by an additional non-significant digit and used as a job number eg: 712.1.

712'...' Primary school

Briefing file: Contents

If the project is divided into individual blocks or clear cut departments, the file may be organised primarily on the basis of blocks or departments. In this example it is assumed that the project is treated as a whole. Each heading, taken from Table 0, represents one section of the file, possibly as little as one item of information. Table 0 numbers can be used to number sections, if required.
Because the sections are in a standard order, every project in the office can be organised on a similar basis to assist the feedback of information from job to job.

Administrative facilities
Headmaster's office

Health, welfare facilities
Medical inspection (MI) room

Educational facilities
School hall including stores
Classrooms including stores
Practical areas
Garden room

Circulation facilities
Entrance halls, circulation space

Rest, work facilities
Staff room

Eating, culinary facilities
Dining area
Kitchen etc

Sanitary facilities
Toilet areas
Changing rooms

Control facilities
Boiler room
Fuel store

External facilities
Patio
Garden
Playing fields

Application 1: Project information

Information for the briefing file may be collected by a formal technique as originally shown in the 1971 Project Manual, or made up using notes of meetings annotated with Table 0 and Table 1 codes in the margin.

Figure 4
Briefing note.

```
(42)   Mr Smith asked that particular attention should be paid to the choice
       of vandal-proof wall finishes to the public areas on the third floor.
       He would need to approve the final selection made.
```

Information on each element can be collated and re-written as necessary to form the outline technical specification, which forms the basis for further work.

Figure 5
Outline technical specification.

```
(42)   INTERNAL WALL FINISHES

       GENERALLY: emulsioned walls, plastered throughout with the following
       exceptions:

       Hall interior: fair faced facing bricks (as exterior - see element
       (21): External walls).

       Cupboards and stores: fair faced blockwork, bagged joints and emulsioned.

       Above practical benches: include for 400mm high formica splashback (on
       18mm blockboard) above all fixed benching.

       Pinboarding: to be 9mm Sundeala on 38 x 9 s.w. battens forming a 600
       grid. Boarding to form continuous strip around classrooms, 1200 wide and
       750 from finished floor level. Include for 25 x 12 painted softwood stop
       bead to top and bottom edges. See detail D(42)1.
```

Application 1: Project information

This may include related design sketches numbered and filed by Table 1.

Figure 6
Design sketch.

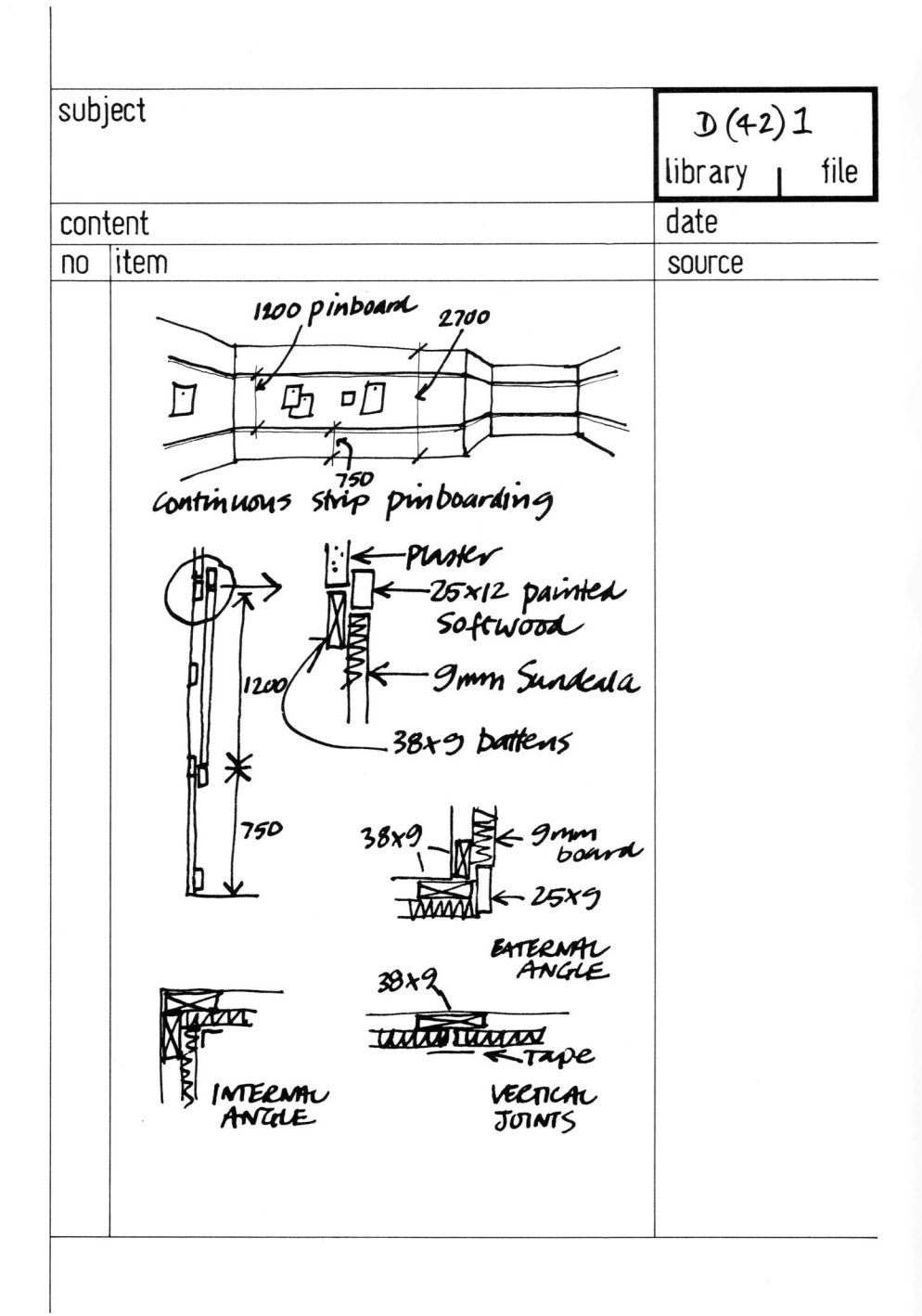

subject		D (42) 1
		library file
content		date
no	item	source

Application 1: Project information

The outline technical specification forms the basis for the initial cost plan, and an edited version of it can be used on or with drawings submitted for building regulations approval.

Figure 7
Initial cost plan.

Code	Element	Specification	Element £	m² £	Total £	m² £
(32)	Internal doors	Flush with glazed panel, formica rails, softwood frames Architraves both sides, fanlights, ironmongery Folding partitions - hall/dining, gym store, chair store Servery openings with roller shutters - kitchen	2980	2.06		
(34)	Balustrades	Hall and library voids - pine rails and hangers	720	0.50		
(37)	Rooflights and glazing	12 No. 1200 x 1200 rooflights Splayed glazing - 6mm wired cast glass in timber frames	1800	1.25		
(39)	Summary				9050	6.26
(42)	Internal wall finishes	Plastering and emulsion paint generally Facing bricks to hall Fair faced blockwork and emulsion paint to cupboards and stores Formica splashbacks to wall benches Pin boarding to bases	4510	3.12		

Application 1: Project information

1.22 Production information

In some cases the whole of the documentation including the drawings is presented in a traditional way up to a very late stage, even until contracts have been let, and then entirely restructured using CI/SfB. This is not advisable. If Table 1 is not used for briefing or for outline technical specification, it should be introduced by the beginning of RIBA Plan of work Stage E. If this is done, architects and consultants will benefit from continuity throughout their detailed work.

Whether or not Table 1 has been used at early design stages, the first step at the beginning of Stage E is the preparation of a drawings programme to identify the drawings required. One method of doing this is described in the RIBA publication Resources control, but alternatively the design team can simply make a provisional list, as Figure 8, of all the drawings likely to be required.

Figure 8
Provisional list of drawings required.

Elements	Drawings and Drawing numbers	Estimated Nos. drwgs	details	Production
(--) COMPLETE PROJECT	Building regulations drawings: (a) Key negatives (b) Schedule of finishes (c) Additional information required for building regulations approval			
(1-) SUBSTRUCTURE	Foundation Plan (1-)10	1	-	Copy negative from Ground Floor Plan with outline of foundations, with foundation sizes and detail numbers marked
(13) Floor bed	Details (13)30 to 31	1	2	Sections. A3 size drawing
(16) Foundations	Details (16)30 to 32	1	3	Sections. A3 size drawing
(2-) PRIMARY ELEMENTS incorporating	Plans (2-)10	1	-	Copy negative of Ground Floor Plan with building sizes and detail numbers marked
(21) External walls	Details (21)30 to 32	1	3	Sections. A3 size drawing
(22) Internal walls	Details (22)30 to 32	1	3	Sections. A3 size drawing
(23) Floors	Details (23)30 to 31	1	2	Sections. A3 size drawing
(27) Roofs	Roof Plan (27)10	1	-	New Drawing - see Key negative
	Details (27)30 to 34	1	5	Sections. A3 size drawing
(28) Frame	Plan (28)10	1	-	Copy negative of Ground Floor Plan with structural sizes and detail numbers marked
	Detail (28)30 to 31	1	2	Sections. A3 size drawing
(3-) SECONDARY ELEMENTS incorporating	Plan (3-)10	1	-	Copy negative of Ground Floor Plan with window and door references and sizes marked
(31) External Openings	Schedule (31)20	1	-	External window and door schedule

Application 1: Project information

This list, with the design drawings, the outline technical specification and any supporting sketches numbered and filed by Table 1, can be used by the production team to develop working drawings as outlined in the next section, through RIBA Plan of Work Stages E and F. At the same time, a provisional list of specification headings and codes can be prepared on the basis of the outline technical specification and set down in terms of the office master specification (see Project information 1.4) as follows:

Figure 9
This illustration is taken from a provisional list of annotations for drawings, incorporating codes based on the National Building Specification.
The list was prepared for use on housing contracts by the Douglas Smith Stimson Partnership in order to ensure consistent terminology and coding in all documents concurrently prepared – drawings, specifications, bills of quantities.
A typical cross reference on the drawings might read '1:3:6 concrete E31/1'.
An identically worded and coded heading would be used in the relevant sections of the specification and bill of quantities.

NBS code	Section title	Drawing annotation	General information
E01	Making concrete		Prepare schedule of types of nominal mixes as described in NBS Section Y31. Normally only three types will be required – one 1:3:6 mix and two 1:2:4 mixes, one not requiring test cubes for slabs etc. and one requiring testing for reinforced work.
E11	Formwork for concrete		Ascertain whether sulphate resisting required – if so additional types will be required.
E21	Reinforcement	-- bars E21 -- fabric E21	State type and or size/weight. Provisional Sum or provisional quantity of reinforcement to be listed to be used in bad ground conditions as directed.
E31	Concreting	typical annotations: 1:3:6 concrete E31/1 1:2:4 concrete E31/2 1:2:4 concrete E31/3	This typical annotation gives unique identification by both words and numbers. When there is more than one type of the same mix, the code alters to distinguish rather than lengthening the description. These are nominal mix types – should structural engineer require design mixes a separate schedule may be prepared. Consider use of trench fill foundations.
E42	Finishes and tolerances for concrete	basic finish E42:31 plain finish E42:32 trowelled/floated finish E42:41 tamped finish E42:43	

The specification can be finalised and the drawings coded and annotated as they are prepared, to provide a complete, systematically-arranged record of decisions affecting construction.

Application 1 : Project information

1.3 Production ('working') drawings
See also Project information 1.1, 1.2

1.31 A simple drawings system

There are two basic approaches to the assembly of views and arrangement of drawings in a set of working drawings. In the arrangement called 'traditional' there are two broad categories – general arrangement drawings and detail drawings. Each shows, at various scales, views either of the whole building or of its parts in plan, elevation or section.

Few conventions relating to traditional drawing are concerned with the placing of information. Characteristically, traditional sets have large sheets which are usually well filled. Titles either broadly identify sheet content (eg door details) or state the type of view (eg elevations). Sheets are numbered consecutively as they are produced; groups of drawings dealing with one subject may be found together and numbered consecutively but there is no guarantee that other drawings concerned with the same subject do not exist elsewhere in the set.

'Structured', 'systematic' or 'coordinated' sets, on the other hand, aim to provide a complete and readily-understood framework for information, with separate drawings for defined subjects.

In order to achieve this as far as is possible within the practical limitations of each project, a good set of drawings should meet the following criteria, identified by Building Process Division, BRS, in a study of 'Working Drawings in Use' carried out in 1972 and 1973 and reported in BRE Digest 172 and, in more detail, in BRE Current Paper 18/73.

Search pattern (division of information between drawings): the set should be designed so that the first group of drawings gives overall views of the project and leads, by means of a simple, natural search pattern, to small sub-groups of detail drawings.

Titles: each drawing should have a short, clear but informative title accompanied by a related drawing number for filing purposes.

Content and cross references: the set should give information on the shape, dimensions, position, composition and fixings of each component part of the building. Direct cross references should be given whenever possible, leading users straight to individual drawings giving more detailed information.

The rest of this section describes one way in which these requirements may be met. Where this needs to be modified for particular office or project circumstances the detailed suggestions made in the Project Manual should provide a useful source of ideas. The system consists basically of three main series of drawings, Figure 10.

Application 1 : Project information

Figure 10
Location drawings showing where work is to be done are unique to each project. 'Kit of parts' assembly and component drawings can often be used or adapted for future projects.

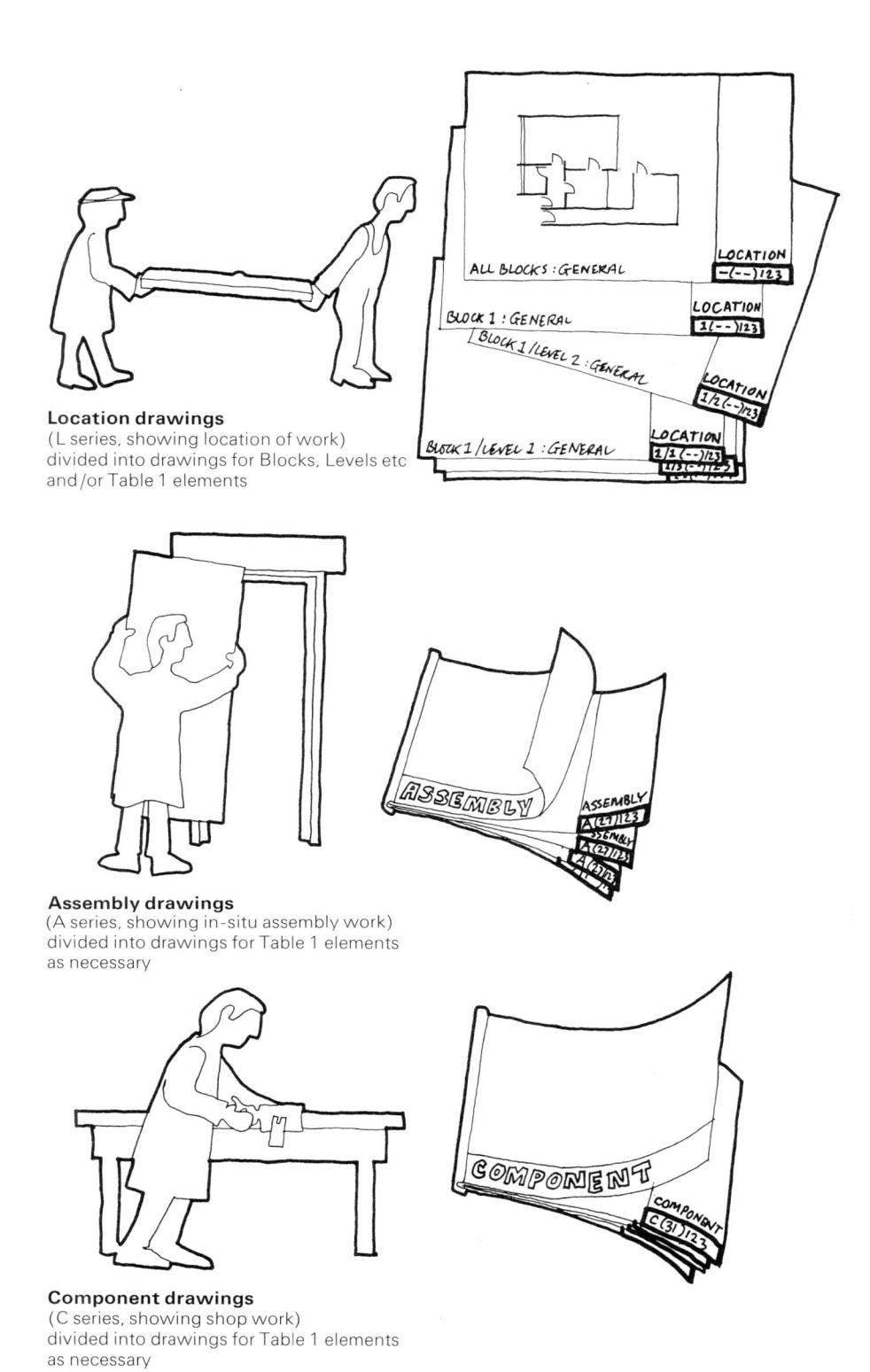

Location drawings
(L series, showing location of work) divided into drawings for Blocks, Levels etc and/or Table 1 elements

Assembly drawings
(A series, showing in-situ assembly work) divided into drawings for Table 1 elements as necessary

Component drawings
(C series, showing shop work) divided into drawings for Table 1 elements as necessary

Application 1: Project information

As a first step, a conventional set of drawings for a previous project can be checked against the detailed recommendations that follow and a revised list of drawing titles and numbers made for it in terms of the system. This may show that the system represents little more than a tidying-up of normal practice in the office and can be introduced without difficulty. General arrangement drawings may be almost unchanged except that they will include many more cross references; (1–) drawings may be similar to traditional foundation and slab layouts; (2–) drawings may resemble traditional setting-out drawings and (3–) and (4–) drawings may be almost identical to normal door, window and finishing schedules.

1.32 Location drawings (L series)

Location drawings show the overall arrangement of the project and the geographical location of work. Examples are plans, elevations, and sections etc of the site; of particular blocks; of external works; of particular departments, floors, or rooms, also schedules of work to openings, finishes etc. It will not usually be possible to re-use location drawings. All location drawings should be on the same paper size, large enough to give an overall view of the project. A1 size is often used.

Division of information between drawings
Drawings are subdivided by scale in the traditional way, starting with 1:100 or smaller scale drawings showing the project as a whole. The first drawing, giving an overall view of the project (like the traditional $\frac{1}{8}''$) may include a brief key to the drawing numbering system giving the meaning of the symbols used on the drawings, see Figure 1. The drawing may also include a brief technical specification for Building Regulations approval purposes (see Project information 1.2).

These small scale drawings lead on to sub-groups of larger scale drawings for particular blocks (including external works), levels, rooms etc.

On small simple projects all location drawings may be general arrangement drawings, showing many or all parts of the project and many or all Table 1 subjects. But if several location drawings are needed for any particular part of the project, it may be convenient to restrict some of them to particular Table 1 subjects.

It may, for example, be useful to take copy negatives of floor plans to show pipe runs on an L(5–) services drawing or even for specific services such as L(52) Drainage layout, but use of the system does not require that this be done, and elaborate copy negative techniques will be best avoided when using CI/SfB for the first time.

On very large projects in which many drawings are to be limited to particular blocks, levels, house types etc, it may be useful to incorporate codes for these things into the filing number but again, this sort of project size and sophistication is best avoided when using CI/SfB for the first time.

Application 1: Project information

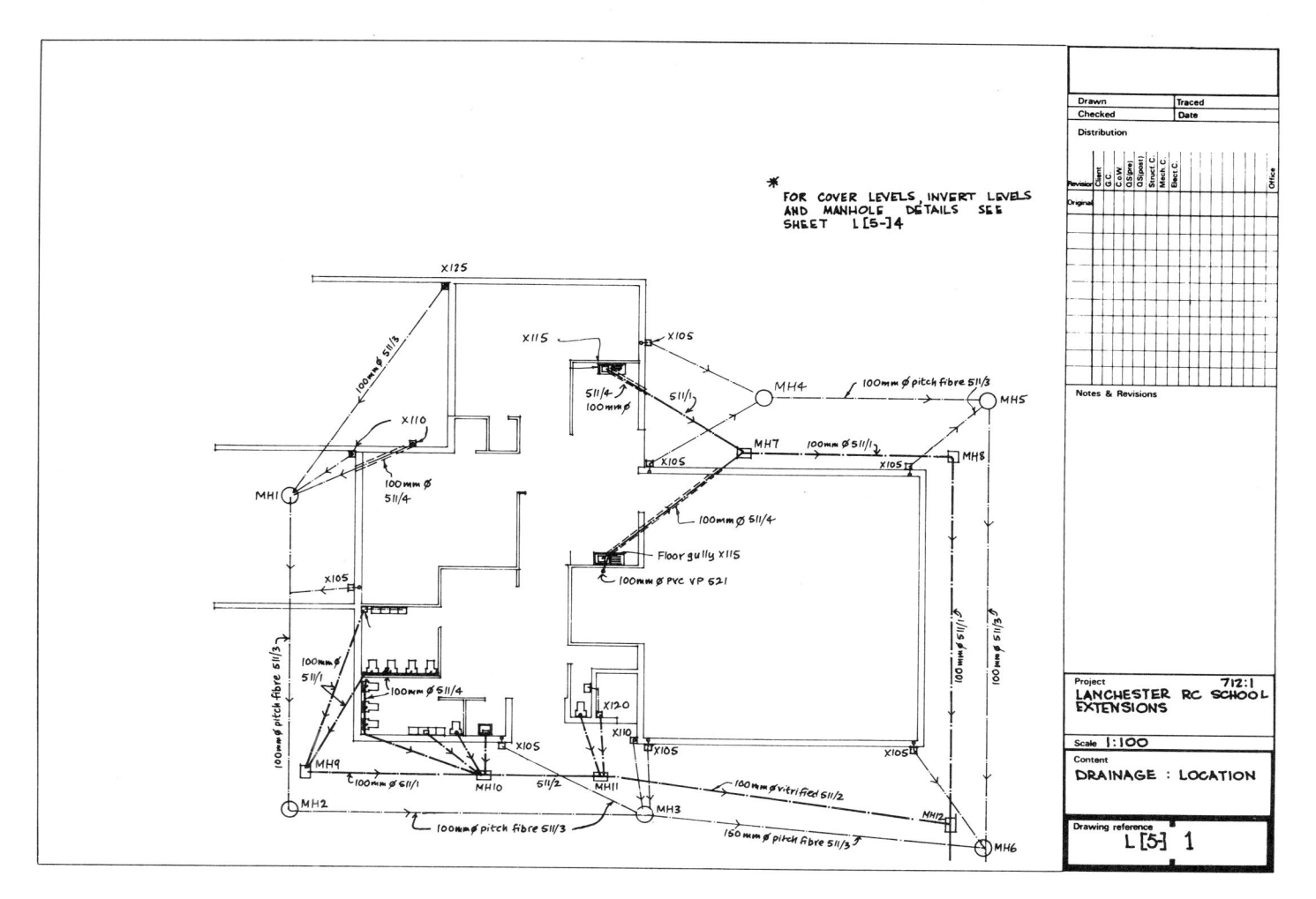

Figure 11
Drainage: location L(5–)1. Size A2.
Could have been numbered L(52) rather than
L(5–), but in view of the small number of
services drawings on this job it was decided to
code them all (5–), viz L(5–)1 Drainage;
L(5–)2 Water supply, heating; L(5–)3
Convector schedule and so on.
The annotations on the drawing refer to the
specification and bill of quantities.
They are based on NBS; see also Figure 9 and
Project Information 1.4.

Application 1: Project information

Titles and drawing numbers
Location drawings which are general arrangement plans, sections, elevations etc, not limited to any particular Table 1 subject, may be numbered L1, L2 etc or L(– –)1, L(– –)2 etc. or L(0)1, L(0)2 etc.

Location drawing showing many or all services

Drawing number is 'L brackets five dash drawing one'

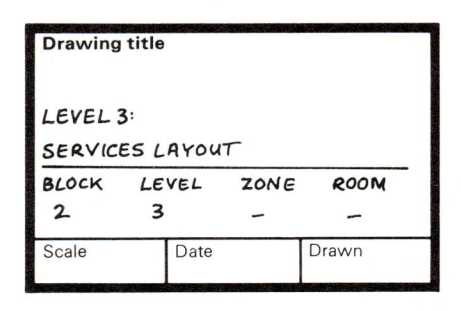

```
Drawing title

LEVEL 3:
SERVICES LAYOUT
BLOCK   LEVEL   ZONE   ROOM
  2       3      –      –
Scale   Date           Drawn
```

```
L (5-)      1
```

Location drawing showing water supply services layout only

Drawing number is 'L brackets five three drawing one'

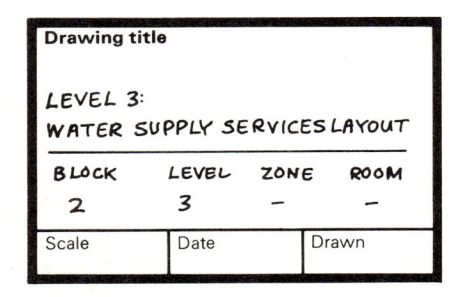

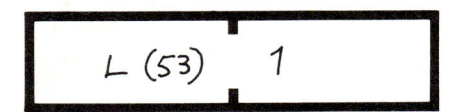

```
Drawing title

LEVEL 3:
WATER SUPPLY SERVICES LAYOUT
BLOCK   LEVEL   ZONE   ROOM
  2       3      –      –
Scale   Date           Drawn
```

```
L (53)      1
```

Part of the title box layout used on RIBA tracing paper is shown in Figure 1a.

Content and cross references
Dimensions, horizontal grids and vertical reference planes sufficient to position all assemblies and components should be given as described in BS 1192, also room names and numbers and numbers for window and door openings. When looking for information, users will frequently consult the location drawings first, so these should show as many cross references to A series drawings as possible. They may also give cross-references to C series drawings and to information on composition and fixing in the specification and bill of quantities (see Project information 1.1, Figure 1c).

Application 1: Project information

1.33 Assembly drawings (A series)

Assembly drawings show fixing (construction) work which is not necessarily limited to one specific location. It should often be possible to re-use much of the information on assembly drawings. A3 size is frequently used but some offices use A1 size for both L and A series drawings.

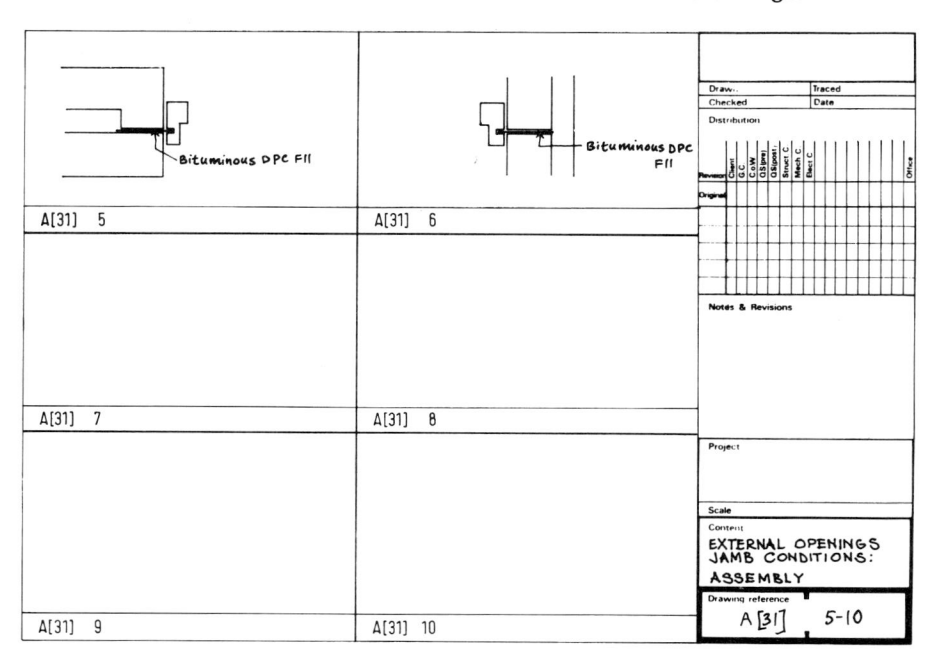

Figure 12
External openings jamb conditions: assembly A(31) 5–10. Size A3.
The annotation 'Bituminous DPC F11' on the drawing refers to the detailed information given in the specification and bill of quantities. It is based on NBS; see also Figure 9 and Project Information 1.4.

Division of information between drawings
Drawings are subdivided by subject, using Table 1 element definitions as a guide to what should be included on each drawing. A series drawings showing elements from **different divisions** in Table 1 should be allocated to the division nearest the beginning of Table 1 eg: heating pipe work ducts in floor beds should be shown on floor bed drawings numbered (13) or on substructure drawings numbered (1–).

When deciding how to deal with junctions between elements **in the same division** of Table 1, eg: ROOF/WALL junctions, producers should consider which element users are most likely to look under, or have a simple rule eg: use the highest code position – in this case (27) Roof:

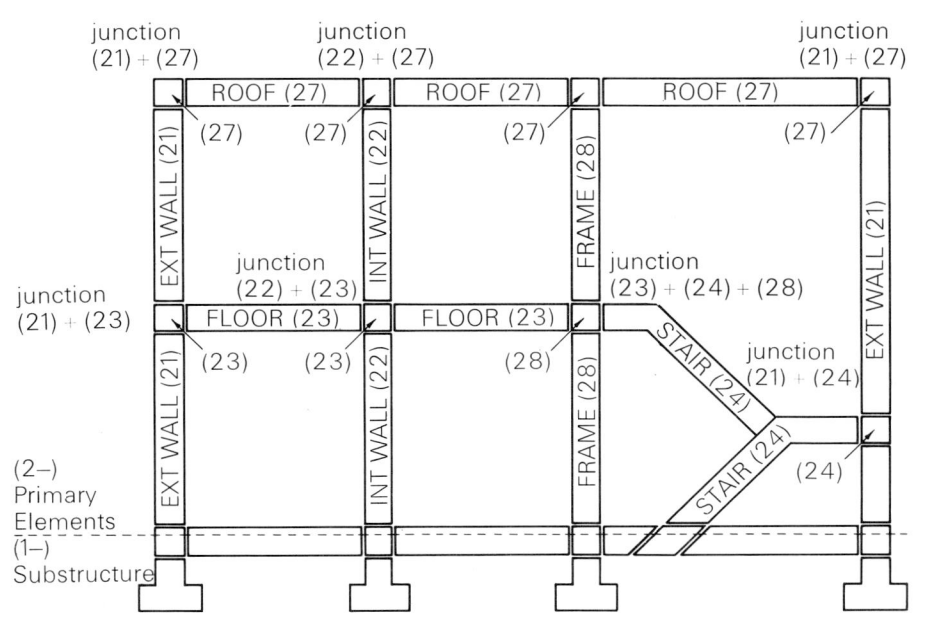

Figure 13
Code precedence for assembly information.

Application 1: Project information

Assembly drawing showing jamb details (No. 5 to 10) for door openings in external walls.

Drawing number is 'A brackets three one drawing five ten.'

Titles and drawing numbers

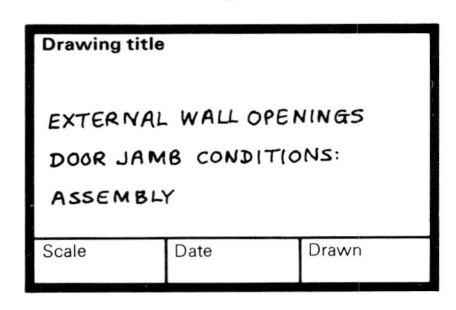

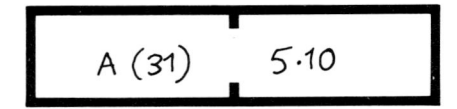

> **Drawing title**
>
> EXTERNAL WALL OPENINGS
> DOOR JAMB CONDITIONS:
> ASSEMBLY
>
> Scale | Date | Drawn

A (31)   5·10

The title in this example does not precisely repeat the relevant Table 1 heading but gives useful additional information. It may sometimes be desirable and possible to indicate as part of the title, the location on the project at which the assembly is to be used.

Content and cross references
Horizontal grids and vertical reference planes may be given on A series drawings, but dimensions that tie them to a particular location in the project should be avoided. Cross references to other A series drawings, to C series drawings and to information on composition and fixing in the specification and bill of quantities should be included (see Project information 1.1, Figure 1c).

1.34 Component drawings (C series)

Component drawings show unfixed components. It should often be possible to re-use component drawings on future projects without much alteration. A3 size is frequently used, but some offices keep to A1 size for L, A and C series.

Figure 14
Boiler house door: component C(31)4.
Size A3.
The annotation 'Softwood H24/1' on the drawing refers to the detailed information given in the specification and bill of quantities. It is based on NBS; see also Figure 9 and Project Information 1.4.

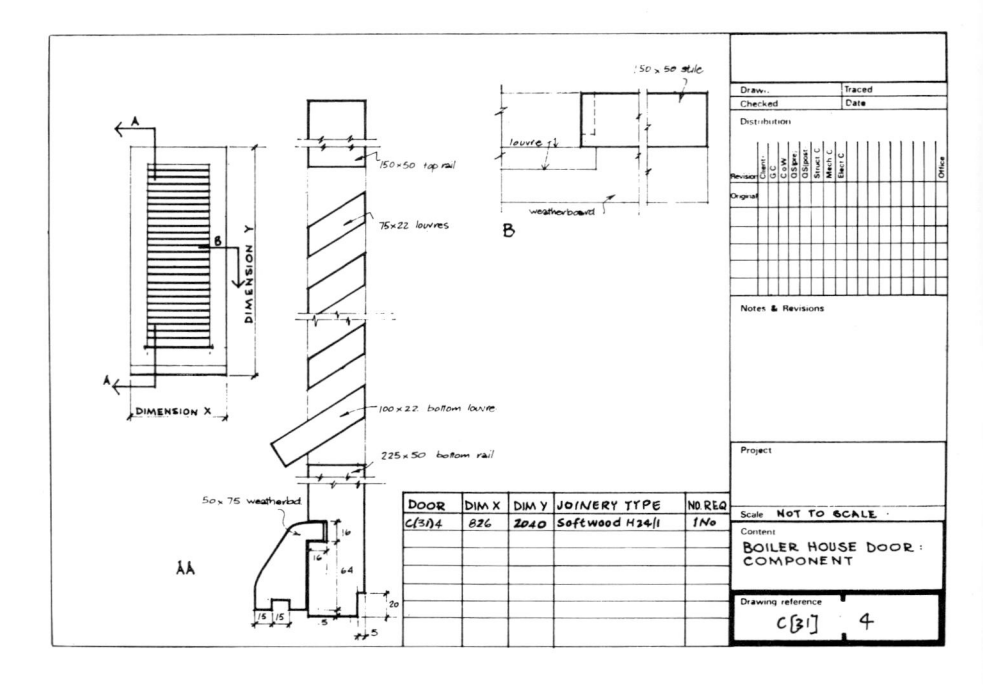

Application 1: Project information

Division of information between drawings
Drawings are subdivided by subject using Table 1. (Any drawings showing very basic products like aluminium sections may be coded, titled and filed using Table 2.)

Titles and drawing numbers

Component drawing showing boiler house door frame.

Drawing number is 'C brackets three one drawing four'.

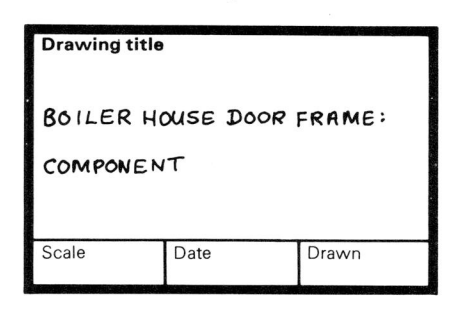

```
Drawing title

BOILER HOUSE DOOR FRAME:
COMPONENT

Scale    Date    Drawn
```

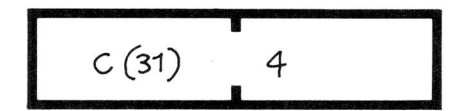

```
C (31)    4
```

The title in this example gives useful specific information. It may sometimes be desirable and possible to indicate as part of the title, the location (as here) or assembly for which the component is to be used.

Content and cross references
Dimensions that tie C series drawings to a particular location or assembly should be avoided. Cross references to other C series drawings and to information on composition in the specification and bill of quantities (see Project information 1.1, Figure 1c) should be included. Cross references to measured items and specification items should normally be put on the most detailed relevant drawing. There is no point in giving the cross reference to specification clauses for the manufacture of a timber window ten times over on a location drawing if it can be given once only on the component drawing for the window.

1.35 Engineering drawings

The drawings system can be applied to structural engineering drawings, with the addition of:

Reinforcement drawings (R series, showing reinforcement)	divided into drawings for LEVELS etc and/or TABLE 1 elements, as necessary, or numbered R1, R2 etc

Note that:
1. Structural engineering drawings can use a simple ad hoc classification system for assembly and component drawings for items applicable to several elements. These are likely to include slabs, columns and beams.
2. In the case of mechanical and electrical engineers' drawings, other special series may be necessary, eg: duct work drawings ('D' series).

Application 1: Project information

1.36 Beware!

The system does not necessitate use of particular drawings sizes or title boxes, particular scales, the metric system, dimensional coordination, grids or copy negatives, the National Building Specification, or elemental bills of quantities. It should not be used first on large complex schemes which would tax any office. It should not lead to the production of large numbers of drawings where a few have always been adequate in the past; or codes without accompanying plain language annotations. Instant economic benefit is possible but not to be expected. The system should be introduced only if it is accepted that it is likely to bring advantages to the office itself as well as to outside users of project documents produced in the office. It should be carefully explained to all those who will use coordinated drawings based on it (see Project information 1.1).

1.37 Analysis of a set of coordinated drawings

The analysis following of a complete coordinated set of drawings used on a live project is not presented as a model to follow — every job has its special requirements — but it illustrates a number of points which need consideration on many projects.

Project description
Two-storey office building and a detached single-storey garage and plant room. Concrete frame, brickwork walls, in-situ concrete floors and flat roofs. Four series of drawings: location, assembly, component and schedules. Each series bound separately and provided with its own contents list; the location drawings being A1 and the rest A3.

The coordinated set of drawings is shown on the following table:

Application 1: Project information

Subject		L series drawings (A1 size)	Location schedules (A3 size)	A series drawings (A3 size)	C series drawings (A3 size)	Totals	
(--)	General arrangement	9				9	9
(21)	External walls			8	3	11	
(22)	Internal walls			5		5	
(24)	Stairs			2		2	22
(27)	Roofs	1		3		4	
(3-)	Secondary elements			2		2	
(30)	Site secondary elements				2	2	
(31)	Ext. walls secondary elements		3	11		14	
(32)	Int. walls secondary elements		3	4	2	9	31
(33)	Floor secondary elements			1		1	
(34)	Stair balustrades			1		1	
(35)	Suspended ceilings			1		1	
(37)	Roof secondary elements			1		1	
(4-)	Finishes				1	1	
(40)	Site finishes	1		7		8	
(42)	Int. wall finishes		1	1		2	16
(43)	Floor finishes		1	1		2	
(44)	Stair finishes			1		1	
(45)	Ceiling finishes		1	1		2	
(50)	Site services	2	2	9		13	16
(52)	Drainage	2			1	3	
(70)	Site fittings			1		1	
(71)	Circulation fittings			1		1	
(72)	Room fittings			5		5	11
(74)	Sanitary fittings		1	1		2	
(76)	Storage fittings			2		2	
		15	12	69	9	105	

Note that:
1. There are no drawings in the (1–) series since the Structural Engineer's drawings (not coordinated with the architect's drawings) cover these elements.
2. Drawings coded (21) and (22) include details of junctions between the walls and the concrete frame. Otherwise the highest number in the (2–) series is used to code junction details, eg: a drawing containing (21) and (27) information is coded (27).
3. The (3–) details include information about the assembly of internal plywood boarding to form ducts by the window openings.
4. Information on external works is split between (30), (40), etc not grouped at (90).
5. The schedules in (31) and (32) are rather isolated. Some of them are not referred to from elsewhere and they do not refer to assembly or component drawings.
6. The main finishes drawings are the schedules in the (4–) series which are referred to from location drawings.

Application 1: Project information

1.4 Specification and quantities

CI/SfB is used for arranging specifications and quantities, whether as two separate documents or as preambles and measured items in a single document.

1.41 Specifications (preambles)

A method of compiling office and project specifications is described in detail in the National Building Specification (see Project information 1.22), and is not repeated here. Both the NBS itself and office or project specifications based on it are arranged largely by CI/SfB Table 2, although knowledge of CI/SfB is not necessary for use of NBS. Cast in-situ work, for example, is divided into the following NBS work sections:

E11 Formwork for concrete
E21 Reinforcement for concrete
E32 Making concrete
E33 Concreting
E43 Finishes and tolerances for concrete
E51 Water excluding concrete construction
E52 Water retaining concrete construction
NBS also includes work sections numbered by Table 1
eg: 522 Drainage pipe work above ground, surface water

CI/SfB can be used to produce a variety of specification formats in addition to the one adopted by NBS.

Figure 15
A page from job specification, based on the 1973 edition of NBS.

```
E21        REINFORCEMENT FOR CONCRETE

           COMMODITIES
E21:H      BARS
   Hh2.02  Tag clearly bundles of reinforcement with bar schedule and bar
           mark references.
   Hh2.10  Hot rolled mild steel bars: to BS 4449.

E21:J      FABRIC
   Jh2.10  Steel fabric: to BS 4483.

E21:X      ACCESSORIES
   Xt6.01  Provide ordinary spacers as necessary to support reinforcement in
           position. Special spacers (e.g. chairs in thick slabs) are shown on
           drawings and schedules.
   Xt6.10  Cover spacers: any of the following types, unless specified
           otherwise:
           1. Concrete: made with 10mm aggregate. In exposed work to match
              surrounding concrete.
           2. Mortar: made with cement sand mortar 1:2.
           3. Plastics: approved type(s).
   Xt6.A   Cover spacers for memorial wall: concrete made with 10mm aggregate
           and matching surrounding concrete.

           WORKMANSHIP
E21.1      GENERALLY
     1051  CLEANLINESS: reinforcement to be clean and free of all loose mill
           scale, loose rust, oil and grease at time of placing concrete.

E21.2      CUTTING AND BENDING
     2051  GENERALLY: cut and bend bars to BS 4466 and to schedule provided,
           unless otherwise instructed.
     2151  REBENDING: do not rebend bars without approval.
     2201  ADJUSTMENTS: provide on site facilities for hand bending to deal
           with minor adjustments.

E21:3      FIXING
     3051  LAPS OR SPLICES: obtain instructions if details are not shown on
           drawings.
     3201  SECURE reinforcement adequately with tying wire, approved steel
           clips, or tack welding if permitted. Bend wire well back clear of
           forms.
```

Application 1: Project information

1.42 Quantities

Figure 16
Elemental (Table 1) bill of quantities, each element subdivided by types of work using Tables 2/3 or NBS.
Talbot and Armstrong, Quantity Surveyors, Durham.

Elemental format

Elemental bills of quantites are arranged using Table 1 headings and codes, so that drawings and quantities follow the same sequence:

		FLOORS	CA	Qty	Unit	Rate	(23) £
		Hi1.2 – Tanalised Structural Softwood					
	a	2" x 7" Floor joists		425	LF		
	b	2" x 9" Floor joists		215	LF		
	c	1" x 9" Strutting in short lengths spiked between joists		12	LF		
	d	Notch 2" softwood for small pipe	(52)(53)(56)	68	No.		
	e	Bore 2" softwood for ¾" bolt		14	No.		
		Rj1.1 – Weyroc Flooring Grade					
	f	18mm Weyroc laid horizontally to joists at 16" centres, fixed with lost headed nails		82	SY		
	g	Hole through Weyroc for small pipe	(52)(53)(56)	32	No.		
	h	Notch Weyroc around newel 8" girth	(34)	4	No.		
		Xt6.1 – Mild Steel					
	i	¾" Diameter bolt 6" long with head nut and two washers		7	No.		
	j	Protection Allow for protecting the work in this Element			Item		
				To Summary			£

In Figure 16 the first column (headed 'CA') is used for cost analysis where the cost allocation of items to services elements necessarily differs from the allocation for construction purposes. For example, holes through flooring are constructionally part of (23) Floor, but are generated by (52) Drainage, (53) Water supply or (56) Heating, so are costed with these elements. Alternatively, builders work items can be included with the services items which generated them, with a cross-reference from the relevant building element.

NBS codes can be included in the measured items to provide direct cross-reference to the relevant sections of the project specification. NBS work sections can also be used as main subdivisions of elements.

Bills of quantities arranged primarily by Table 1 elements are convenient for design cost control and for the contractor's production staff. Some other possible arrangements are listed and illustrated in Figures 17 to 21.

Application 1: Project information

SMM or Scottish Trades format (Figure 17)
Each SMM or Scottish trades work section can be subdivided by Table 1 to assist cross reference from the bill to the drawings. Relevant Table 1 codes can be given in the top right hand corner of each page to help users searching the bill for information on a particular element. Individual items can cross-refer to the specification or preambles.

Multiple format (Figure 18)
Bills of quantities which restrict each page to information common to one Table 1 element and one NBS work section have the slight disadvantage that they require a good deal of paper but the major advantage that they can be quickly and easily sorted into SMM or NBS order for estimators, and Table 1 order for site staff, without use of computers.

Labour material format (Figure 19)
Bills of quantities have been produced which separate measured items for materials from measured items for labour, and arrange these two parts of the bill in different ways using SfB.

Figure 17
SMM or Scottish trades bill of quantities, each trade subdivided by elements (Table 1).
Armour and Partners, Quantity Surveyors, Glasgow.

				ECB(23),(24)
	EXCAVATION, CONCRETE AND BRICK			
	(23) Floors			
68	150mm Bison or equal prestressed precast concrete floor slabs in clear spans of 4.32 metres between brick supports.	m^2	12	
	Extra for:			
69	finished ends of units.	m	5	
70	solid ends at wall supports.	m	5	
71	Concrete Quality 'C' in reinforced in-situ floor 150mm thick.	m^2	2	
72	Softwood sawn formwork to horizontal soffits and afterwards removing.	m^2	1	
73	Raking cutting formwork.	m	2	
	Reinforcement			
	Note: An item is included in Bill No. 2, Provisional and Prime Cost Sums, for the Supply and Delivery of all Reinforcement.			
74	Taking delivery on site from Nominated Reinforcement Supplier and fixing mild steel fabric reinforcement in floor weighing 0.5 kgs/m^2 including tying wire, distance blocks and ordinary spacers (Provisional Quantity)	m^2	4	
	(24) Stairs			
75	Precast concrete steps overall size 300x182.5mm and 1.175 metres splayed on face and reinforced with suitable handling reinforcement finished on tread, undercut riser and one end with granolithic bedded and jointed in mortar on brick walls.	No	18	
76	Concrete Quality 'C' in reinforced landing 150mm thick.	m^2	3	
		To Collection	£	

Application 1: Project information

Figure 18
Bill of quantities for arrangement by elements (Table 1), types of work (Table 2) or SMM. The use of alternative page numbers eg 'e' element, 'cf/m' construction form/material, assists rapid collation.
Roger Thorpe, Architect, Sheffield.

		(23) FLOORS Eq4 CAST IN SITU CONCRETE	
	(23)Eq4; IN-SITU CONCRETE; PLAIN		
	(23)Eq4 001; Normal; mix 3000lb; ¾" aggregate		
	Suspended floors or the like; horizontal; filling between precast floor units and perimeter walls; not exceeding 12" wide; over soffit units provided by others		
a	5" thick	26 sy	
	(23)Eq4; IN-SITU CONCRETE; REINFORCED		
	(23)Eq4 001; Normal; mix 3000lb; ¾" aggregate		
	Suspended floors or the like; horizontal		
b	6" thick	10 sy	
	Casings to steel beams; horizontal or sloping not exceeding 15 degrees from horizontal		
c	sectional area over 48 not exceeding 144 square inches	1 cy	
	(23) Eq4; FORMWORK TO REINFORCED IN-SITU CONCRETE		
	Formwork generally		
	Soffits; horizontal		
d	floors or the like	9 sy	
	Sides and soffits		
e	beams or the like; horizontal	10 sy	
	(23)Eq4 001; SUNDRIES		
	Holes for pipes or the like; extra large		
f	6" concrete	3 no	
	e32 cf/m 7	To Collection	£

Figure 19
Bill of quantities in two main sections. Labour items are arranged by element (Table 1) subdivided by types of work (Table 2). Materials are arranged by type of work (Table 2) subdivided by material (Table 3). Mulcahy, McDonagh and Partners, Quantity Surveyors, Dublin.

	LABOUR		Qty	Rate	Total
	Activity number	(23) Floors, Galleries			
		E Concreting			
		Labour placing concrete 150mm thick on formwork and vibrating around reinforcement in suspended floors. (Drawings (2-), (23); Specification section E21)			
a	39	1st Floor	150m²		
b	48	2nd Floor	100m²		
c	57	3rd Floor	100m²		
	COMMODITIES				
	Materials code				
a		E Concreting			
	p1	Coarse aggregate as Specification section E31: Ep1	225t		
		(13) 36t (16) 30t (21) 75t (22) 15t (23) 51t (27) 18t			
	p1	Fine aggregate as Specification section E31: Ep1	150t		
		(13) 24t (16) 20t (21) 50t (22) 10t (23) 34t (27) 12t			
	q2	Portland cement as Specification section E31: Eq2	75t		
		(13) 12t (16) 10t (21) 25t (22) 5t (23) 17t (27) 6t			

Application 1 : Project information

CBC formats (Figure 20)
Coordinated Building Communications Ltd (CBC) produce computer bills of quantities which arrange resources by Table 3.

NBS format (Figure 21)
Use of NBS format for bills of quantities results in an arrangement broadly similar to SMM and puts quantities in the same sequence as the specification sections. Where this is done, NBS section codes put on the drawings facilitate reference to the quantities as well as to the specification. Like SMM arranged bills, NBS arranged bills can be subdivided by Table 1.

Figure 20
Computer produced bill of quantities.
The bill shows the quantities required at each floor level in each block.
CBC Ltd, Quantity Surveyors, Copenhagen.

```
ESBJERG CENTRAL HOSPITAL.                    JOB NR. 9101 PAGE   1
CONCRETE WORKS.                              SUB NR. 0005
                                                       75-07-11
-------------------------------------------------------------------
                                        QUANT. UNT   RATE    TOTAL

(23)E  .0001 FLOOR SLABS: CAST I-S CONSTR.  )

        OPERATIONS:

  05.1207 CUTTING, BENDING, PLACING AND    )
          BINDING OF MILD STEEL BAR REIN-  )
          FORCEMENT IN THE FOLLOWING DI-   )
          AMETERS:

  05.1210 DIAMETER 10 MM.                  )

          BLOCK 01 FLOOR 00            11570.0 KG
          BLOCK 01 FLOOR 01             5300.0 KG
          BLOCK 02 FLOOR 02             7102.0 KG

                                    TOT 23972.0 KG

  05.1216 DIAMETER 16 MM.                  )

          BLOCK 01 FLOOR 00             7563.0 KG


  05.2516 PLACING AND VIBRATING OF CON-    )
          CRETE IN SUSPENDED SLAB 200 MM   )
          THICK.                           )

          BLOCK 01 FLOOR 02               63.1 M2
          BLOCK 01 FLOOR 03               21.0 M2

                                    TOT   84.1 M2

        MATERIALS:

  H2.1110 SUPPLYING OF MILD STEEL BAR      )
          REINFORCEMENT, DIAMETER 10 MM.   ) 23972.0 KG

  H2.1116 SUPPLYING OF MILD STEEL BAR      )
          REINFORCEMENT, DIAMETER 16 MM.   )  7563.0 KG

  Q4.1560 SUPPLYING OF CONCRETE 31 N/MM2.  )    17.0 M3
```

Application 1 : Project information

Figure 21
Bill of quantities in NBS order, to parallel NBS based job specification.
Drawings and specification by Faulkner-Brown, Hendy, Watkinson, Stonor, Architects, Newcastle upon Tyne.
Bill of quantities by Gleeds, Quantity Surveyors, Newcastle upon Tyne.

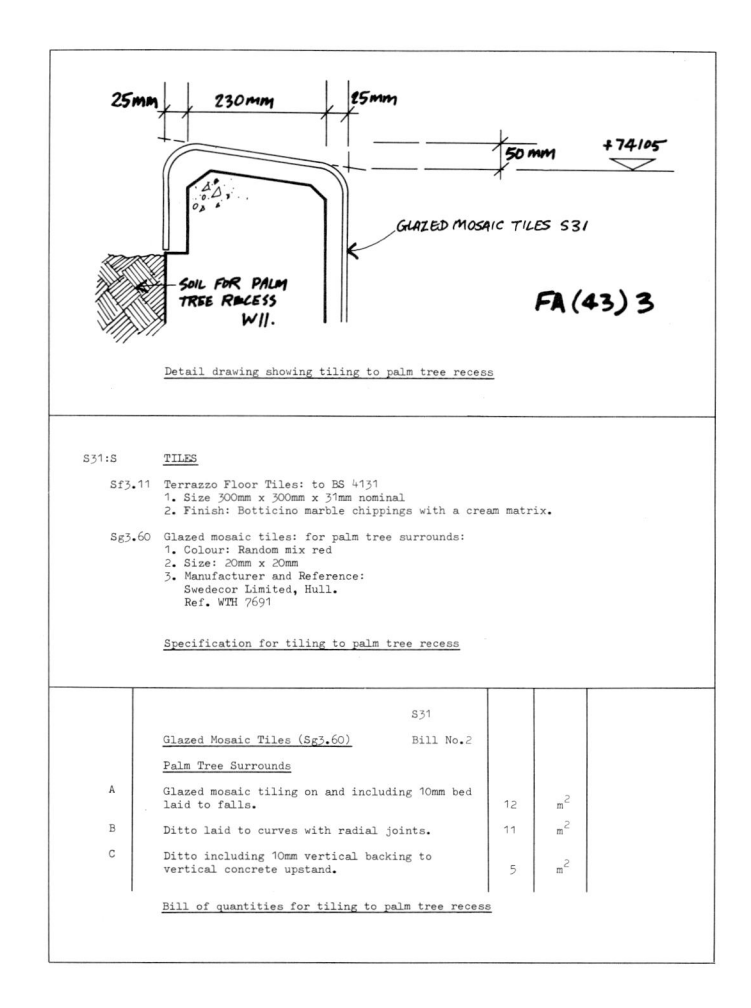

```
Detail drawing showing tiling to palm tree recess
```

```
S31:S    TILES

Sf3.11   Terrazzo Floor Tiles: to BS 4131
         1. Size 300mm x 300mm x 31mm nominal
         2. Finish: Botticino marble chippings with a cream matrix.

Sg3.60   Glazed mosaic tiles: for palm tree surrounds:
         1. Colour: Random mix red
         2. Size: 20mm x 20mm
         3. Manufacturer and Reference:
            Swedecor Limited, Hull.
            Ref. WTH 7691

Specification for tiling to palm tree recess
```

	S31 Glazed Mosaic Tiles (Sg3.60) Bill No.2		
	Palm Tree Surrounds		
A	Glazed mosaic tiling on and including 10mm bed laid to falls.	12	m²
B	Ditto laid to curves with radial joints.	11	m²
C	Ditto including 10mm vertical backing to vertical concrete upstand.	5	m²

Bill of quantities for tiling to palm tree recess

A matrix or list can be included at the beginning of particular bills of quantities to help estimators familiar with SMM to find information quickly.

Figure 22
An SMM arranged contents list to an NBS arranged bill of quantities.

PLASTERWORK AND OTHER FLOOR WALL AND CEILING FINISHES		PAINTING AND DECORATING	
		Internal work	
Plasterboard and skim	R71, P11/1	Prepare and paint 2 coats of emulsion	V52/1
Two coat plaster	P11/2	Prepare, prime, one undercoat and one coat gloss on steelwork	V52/2
Angle bead	P11		
Glazed ceramic tiles	S32		
Vinyl asbestos tiles	T32		
Quarry tiles	S31	Prepare and prime only woodwork	V52/3
Cement and sand beds	P13		
		Knot, prime, stop, one undercoat and one coat gloss on woodwork	V52/4
GLAZING		External work	
Clear sheet glass	R21	Prepare, prime, two undercoats and one coat gloss on metalwork	V52/5
Georgian wired glass	R21		
		Knot, prime, stop, two undercoats and one coat gloss on woodwork	V52/6

Application 1: Project information

These examples show that considerable flexibility is possible, with use of SfB tailored to different circumstances. There is no standard practice in the placing of PC sums etc, but in elemental bills, the preambles or specification is usually treated as a completely separate section of the bill. Preliminaries including contingencies can start the code sequence at (0–); PC and provisional sums can be included in the body of each section or listed at the end of it; dayworks and insurances can be included at the end of the bill at the summary, coded (– –) (Note 1.1) if required, with a reference back to (0–) for pricing purposes.

1.5 Other project information applications

Cost planning by CI/SfB will be particularly advantageous where Table 1 elements are also to be used to organise detailed design drawings and working drawings (Project information 1.1). In Ireland, Table 1 is used for the list of National Standard Building Elements (Note 1.2).

Part of an initial cost plan arranged by Table 1 is shown in Project information 1.1. This provides an important design aid related to the outline technical specification, and also facilitates calculations for costs-in-use and cost checking during the design stage. Final cost plans can be presented, where necessary, in terms of the 'Standard Form of Cost Analysis' (SFCA) (Note 1.3) or as required by Government Departments or other authorities.

The following table will be found useful by those familiar with SFCA who are involved in the preparation of Table 1 initial cost plans. It shows, incidentally, how similar are the definitions of SFCA and CI/SfB elements. For most, only the codes and the order differ.

1.51 Cost planning

1.1   (– –) can be used to summarize either to the front or the end of any collection of information.

1.2   'National Standard Building Elements and Design Cost Control Procedures' available from An Foras Forbartha Teoranta, St Martin's House, Waterloo Road, Dublin 4.

1.3   'Standard Form of Cost Analysis: Principles, Instructions and Definitions', available from Building Cost Information Service, 85–7 Clarence Street, Kingston-upon-Thames, Surrey KT1 1RB.

Application 1: Project information

	SFCA to Table 1
Standard Form of Cost Analysis	Notes for use in preparing Table 1 cost plans to relate to drawings and outline specification

Building substructure: Total SFCA division 1 = Total Table 1 division (1–)

1 Substructure			Define as SFCA, use code (1–)

Building fabric: Total SFCA divisions 2, 3 = Total Table 1 divisions (2–)+(3–)+(4–)

2 Superstructure	2A	Frame	Define as SFCA, use code (28)
	2B	Upper Floors	Define as SFCA, use code (23)
	2C	Roof	
	2C1	Roof structure	Define as SFCA, use code (27)
	2C2	Roof Coverings	Define as SFCA, use code (47) or keep with (27)
	2C3	Roof drainage	Define as SFCA, use code (27)
	2C4	Roof lights	Define as SFCA, use code (37) or keep with (27)
	2D	Stairs	
	2D1	Stair structure	Define as SFCA, use code (24)
	2D2	Stair finishes	Define as SFCA, use code (44) or keep with (24)
	2D3	Stair balustrades and handrails	Define as SFCA, use code (34) or keep with (24)
	2E	External walls	Define as SFCA, use code (21)
	2F	Windows and external doors	Define as SFCA, use code (31), subdivide if necessary
	2F1	Windows	
	2F2	External doors	
	2G	Internal walls and partitions	Define as SFCA, use code (22)
	2H	Internal doors	Define as SFCA, use code (32)
3 Internal Finishes	3A	Wall finishes	Define as SFCA, use code (42)
	3B	Floor finishes	Define as SFCA, use code (43)
	3C	Ceiling finishes	
	3C1	Finishes	Define as SFCA, use code (45)
	3C2	Suspended ceilings	Define as SFCA, use code (35) or keep with (45)

Services and fittings: Note that total SFCA divisions 4+5=Total Table 1 divisions (5–)+(6–)+(7–)+(8–)

4 Fittings and furnishings	4A1	Fittings, fixtures and furniture	Define as SFCA, use code (7–), subdivide if necessary
	4A2	Soft furnishings	
	4A3	Works of art	
	4A4	Equipment	
5 Services	5A	Sanitary appliances	Define as SFCA, use code (74)
	5B	Services equipment	Define as SFCA and use code (7–) **or** define as Table 1 and use detailed codes (73) to (78)
	5C	Disposal installations	Define as SFCA, use code (52), subdivide if necessary. Code (51) may be used for refuse disposal and (52) for drainage only
	5C1	Internal drainage	
	5C2	Refuse disposal	
	5D	Water installations	Define as SFCA, use code (53), subdivide if necessary
	5D1	Mains supply	
	5D2	Cold water service	
	5D3	Hot water service	
	5D4	Steam and condensate	

Application 1: Project information

SFCA to Table 1 continued

5 Services continued	5E	Heat source	Define as SFCA, use code (58) unless heat source is for eg: space heating only, then use code (56)
	5F	Space heating and air treatment	
		Heating only (5F1 to 5F5)	Define as SFCA, use code (56), subdivide if necessary
		Air treatment (5F6 to 5F9)	Define as SFCA, use code (57), subdivide if necessary
	5G	Ventilating system	Define as SFCA, use code (57)
	5H	Electrical installations	
	5H1	Electric source and main	Define as SFCA, use code (61)
	5H2	Electric power supplies	Define as SFCA, use code (62)
	5H3	Electric lighting	Define as SFCA, use code (63), subdivide if necessary
	5H4	Electric lighting fittings	
	5I	Gas installations	Define as SFCA, use code (54)
	5J	Lift and conveyor installations	Define as SFCA, use code (66), subdivide if necessary
	5J1	Lifts and hoists	
	5J2	Escalators	
	5J3	Conveyors	
	5K	Protective installations	Define as SFCA, use code (68), subdivide if necessary
	5K1	Sprinkler installations	
	5K2	Fire fighting installations	
	5K3	Lightning protection	
	5L	Communication installation	Define as SFCA. Use code (64), subdivide if necessary. Note that SFCA includes fire and theft warning installations here but Table 1 normally includes them at code (68)
	5M	Special installations	Define as SFCA and keep together at code (68), **or** allocate as follows: Vacuum cleaning to code (52) Heated water to code (53) Gases supply to code (54) Refrigerated stores to code (55) Travelling cradles to code (66)
	5N	Builders work in connection with services	Define as SFCA and allocate to each service **or** keep together at codes (59) and/or (69) as appropriate
	5O	Builders profit and attendance on services	Define as SFCA and allocate to each service **or** keep together at codes (59) and/or (69), as appropriate

External works: Note that total SFCA division 6 = Total Table 1 division (9–)

6 External works	6A	Site works	
	6A1	Site preparation	Define as SFCA, use code (90.1) or (91)
	6A2	Surface treatment	Define as SFCA, use code (90.4) or (94)
	6A3	Site enclosure and division	Define as SFCA, use code (90.3) or (93)
	6A4	Fittings and furniture	Define as SFCA, use code (90.7) or (97)
	6B	Drainage	Define as SFCA, use code (90.5) or (95)
	6C	External services	Define as SFCA, use subdivisions of codes (90.5)/(90.6) or (95/96) as required
	6D	Minor building work	Define as SFCA, use code (90.2) or (92)

Application 1: Project information

Table 1

Table 1 Form of cost analysis

Table 1 cost plans can be produced in various formats. The following example provides economically for the SFCA cost elements, as shown in square brackets. Indirect costs may be included at each element individually or collected at the '0' or '9' positions:

				£	£
0	**Preliminaries**	(0–) Preliminaries			xxxx
1	**Substructure**	(19) Summary for substructure [1]			xxxx
2	**Structure**	(21) External walls [2E]		xxxx	
		(22) Internal walls [2G]		xxxx	
		(23) Floors [2B]		xxxx	
		(24) Stairs [2D]		xxxx	
		(27) Roofs [2C]		xxxx	
		(28) Frame [2A]		xxxx	
		(29) Summary for structure			xxxx
3	**Secondary elements for walls**	(31) External wall openings [2F]		xxxx	
		(32) Internal wall openings [2H]		xxxx	
		(39) Summary for secondary elements in walls			xxxx
4	**Finishes**	(42) Wall finishes internal [3A]		xxxx	
		(43) Floor finishes [3B]		xxxx	
		(45) Ceiling finishes [3C]		xxxx	
		(49) Summary for finishes			xxxx
5	**Services, mainly piped**	(52) Disposal [5C, 5M part]		xxxx	
		(53) Water supply [5D, 5M part]		xxxx	
		(54) Gases supply [5I, 5M part]		xxxx	
		(55) Space cooling [5M, part]		xxxx	
		(56) Space heating [5F1 to 5F5]		xxxx	
		(57) Air conditioning and ventilation [5F6 to 5F9, 5G]		xxxx	
		(58) Heat source [5E]		xxxx	
		(59) Summary for piped services			xxxx
		ADD for Builders work and attendance			xxxx
6	**Services, mainly electrical**	(61) Electrical supply [5H1]		xxxx	
		(62) Power services [5H2]		xxxx	
		(63) Lighting services [5H3, 5H4]		xxxx	
		(64) Communications services [5L, not alarms]		xxxx	
		(66) Transport services [5J, 5M part]		xxxx	
		(68) Security, control services [5K plus alarms]		xxxx	
		(69) Summary for electrical services			xxxx
		ADD for Builders work and attendance			xxxx
7	**Fittings**	(7–) Fittings [4, 5A, 5B]		xxxx	
		(74) Sanitary fittings [5A, 5B]		xxxx	
		(79) Summary for fittings			xxxx
9	**External works**	(90) External works [6]			xxxx
		(– –) Summary for project			**xxxx**

Application 1: Project information

1.52 Correspondence

The examples of file divisions given below assume that letters from any one correspondent are kept together to avoid the need for extensive photocopying of parts of letters dealing with different subjects. The correspondent is classified, not the letter. Correspondence and minutes of meetings which cannot readily be allocated to any one subject heading can be kept at (A1) and in any case where two file divisions seem appropriate, the one nearer the beginning of the list will normally be used.

(A0) General
eg: files for:
Agreements with clients, consultants and colleagues, project procedures (eg: this Manual, Job Book), fees, and all matters covered by the office administrative files so far as they apply to definitions of responsibility, allocation of tasks etc for the project. Form of contract, when available. Confidential matters.

(A1) General project correspondence, minutes etc
eg: files for:
Client, users and their advisers. Design team consultants. General contractor.

(A2) External controls correspondence, minutes etc
eg: files for:
Consents and approvals, correspondence and minutes of meetings with, for example:
Town planners, ministries, district surveyor, building inspector, fire officer. Adjoining owners. Water Board, Electricity Board, etc.

(A3) Design, design control correspondence, minutes etc
eg: files for:
Time sheets for design. Research for design. Site investigation, factual site data, survey for design. Brief, activity data sheets for design. Architects instructions, variation orders.

(A4) Finance, financial control correspondence, minutes etc
eg: files for:
Job costing. Cost planning if not in QS file. Cost analysis, if not in QS file. Contract procedure, tenders, contract. Valuations, certificates, financial running statements. Draft final account if not in QS file.

(A5) Production, production control correspondence, minutes etc
eg: files for:
Programming. Site meetings, minutes. Site timesheets.

(A6) Vacant

(A7) Quality, quality control correspondence, minutes etc
eg: files for:
Clerk of works' reports. Site architect's reports.

(A8) Handover, feedback, appraisal correspondence, minutes etc

This list of the most usual file types can be followed by one or more files for enquiries and quotations based on Tables 1, 2, 3 and 4 as required.

Application 1: Project information

1.53 Feedback

Systematic feedback from the experience of one project to the next is rare in the building industry, although widely regarded as important. It is much easier to achieve it using structured project information than with traditional information. There are two forms of feedback: the re-use of the project documents themselves (or of copies of them), and the extraction of experience from them. Both can be included in the library as the office's collective memory. Figure 23 shows how re-use headings and codes for A and C series drawings can differ from codes and headings used while projects are live.

Figure 23
The concepts which may be useful for searching in a drawings library are not all suitable for coding a drawing for project use.

```
Project code C(32)45              Library code (32.5)i4(K)123

C       Component drawing

(32)    Doors (Table 1)           (32.5)  Doors sliding (Table 1)
                                  i4      Timber, plywood (Table 3)
                                  (K)     Fire stop (Table 4)
45      Identification number     123     Identification number
```

For architects, the amendment of the office master specification in the light of experience with successive project specifications is one obvious example of feedback. Product Data sheets can be marked to show the experience of using particular proprietary products. Record prints of A and C series drawings can be marked up and filed in the library to show what modifications were needed on site, and why; and user appraisal notes can be added. A copy of the brief, outline specification, cost plan and final account details can be filed in a project file under the appropriate building type. Quantity surveyors can use priced bills of quantities in Table 1 format to feed price records also based on Table 1.

2 General information

2.1 Introduction

General information has been defined as 'Information not particular to one project, but applicable to any project and available to everybody'. It includes 'ex-project' information which is fed back into the office through the library or by other means. The management of general information usually involves the classification, filing, indexing and retrieval of complete documents, most of which will not have been structured by CI/SfB. Each of these aspects is considered in turn.

2.11 How to classify

Each item of information (Note 2.1), unless produced within a particular CI/SfB definition (see eg: Project information 1.2, 1.3, 1.4) needs to be classified and coded by the user if it is to be systematically filed and found when required. Inevitably, many information items deal with compound subjects which fall into two or more of the tables. 'Hospitals' are a type of **building** (Table 0) and 'floor finishes' are **elements** (Table 1).

In which table is a document to be placed if it deals with both subjects? Before classifying an item dealing with 'hospital floor finishes', the classifier must decide whether, for his purposes, he wants items dealing with both buildings and elements to be classified primarily by the **building** category or the **element** category. In other words, he must fix the order of priority (citation order) between buildings and elements. The important thing is to

---

2.1   'Item of information' includes a complete document (eg: book, catalogue, set of drawings); a part document (eg: an individual drawing in a set, a section of a catalogue, a chapter in a book); or an individual clause, paragraph or detail. Each of these three levels can be classified by CI/SfB.

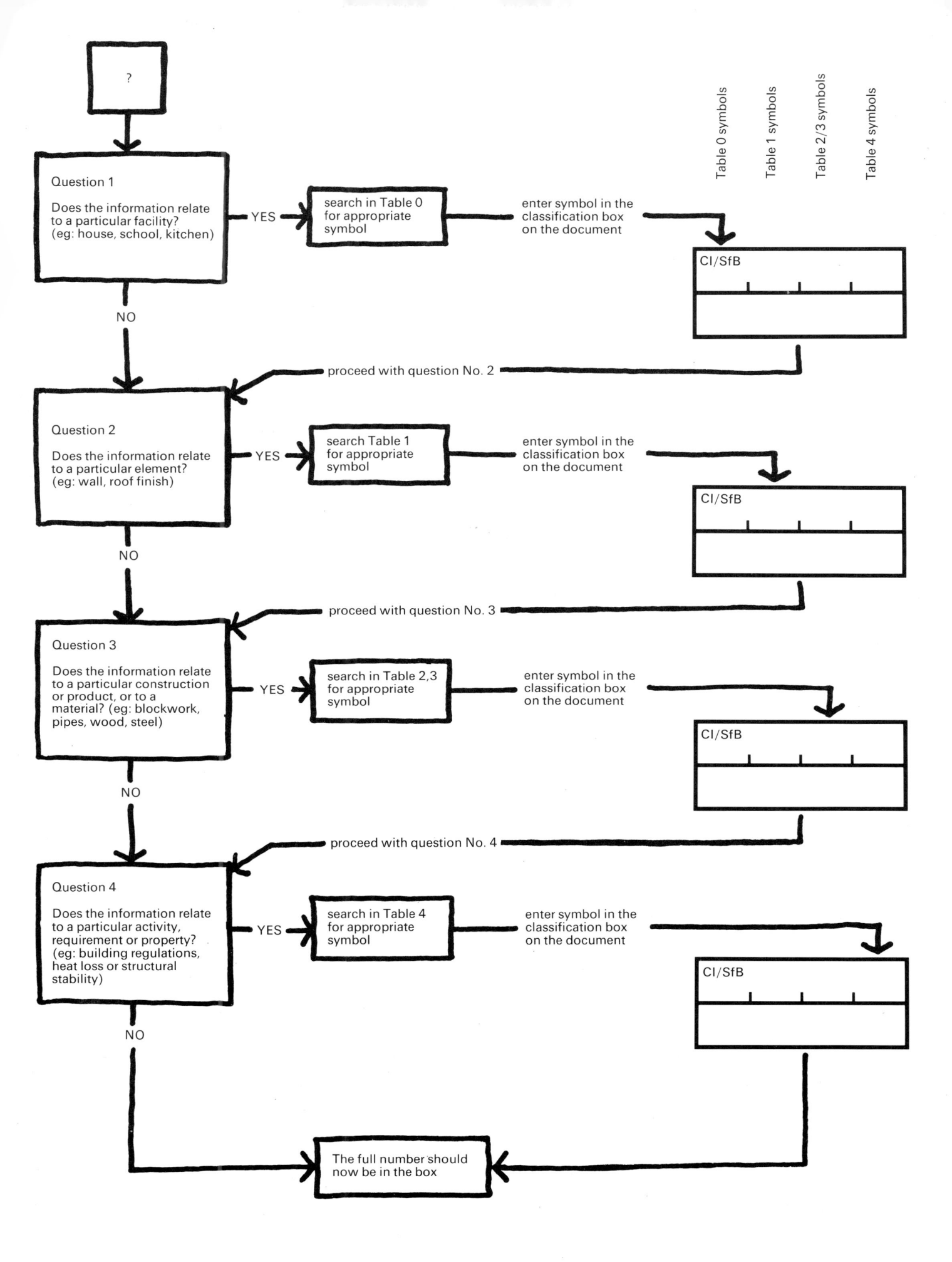

Figure 24
Classification flow chart.

Application 2: General information

have a rule which will deal consistently with all combinations of subjects likely to be met within any one application of the system.

As an example, in most office libraries based on CI/SfB, the classifier asks himself a series of questions, taking each Table 0 to 4 in turn using this order of tables as the citation order.

Thus, for a document on **clay brickwork in external walls.**

Table 0	1. Does it deal with a particular **built environment** in Table 0? Answer: **No.**

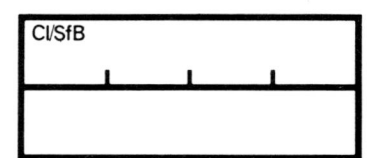

Table 1	2. Does it deal with a particular **element** in Table 1? Answer: **Yes,** External Walls Symbols from Table 1 are (21).

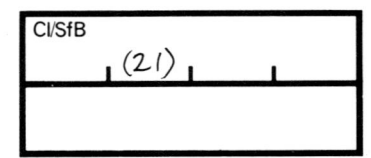

Table 2/3	3. Does it deal with a particular **construction form and/or material** in Tables 2 and 3? Answer: **Yes,** Clay brickwork Symbols from Tables 2 and 3 are Fg2.

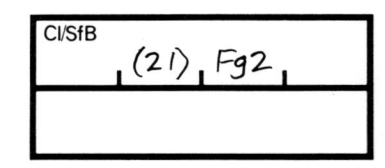

Table 4	4. Does it deal with a particular **activity** or **requirement** in Table 4? Answer: **No.**

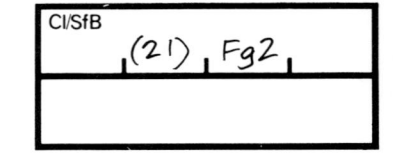

(21) Fg2 is the full reference for the document.

Application 2: General information

A warning

Using this procedure, a document about 'baths for health buildings' may be classified:

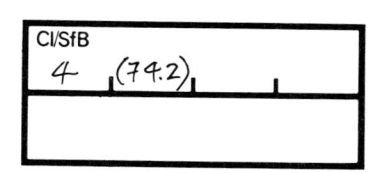

CI/SfB	
4 (74.2)	

but a careful reading may suggest that the references to health buildings are hardly likely to be relevant to most enquiries and that the document would be better classified simply as 'baths':

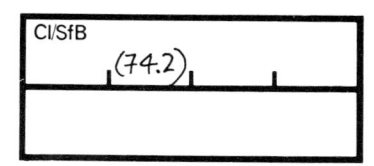

CI/SfB	
(74.2)	

In these examples the codes have been shown in the classification box recommended for trade literature, (see General information 2.2) but they can also be shown without a box round them, eg:

4(74.2)

This form has been used for the remaining examples.

Two subjects from the same table
A document about 'kitchens in hospitals' covers two subjects, 'kitchens' and 'hospitals', from Table 0. Again, an order of priority between them can be decided. The order in which subjects appear may be used as the citation order, giving preference to the subject which comes first in the table. If this is done, such a document will be coded: 41:93.

For many purposes, where this citation order is well established, it will be sufficient to use only the first part of a two-part code from the same table; for this example just: 41.

Two subjects within the same main heading
A document about 'independent secondary schools' covers two subjects, 'secondary schools' and 'independent schools', both from class 71 in Table 0. The first is at code 713, the second at 718. The order of priority between the two possibilities can again be decided according to the order in which subjects appear in the class. Giving preference to the first, and using only the first code, the document will be coded: 713.

Freedom to choose
In all these examples priority has been given to whichever subjects appear first in the tables. This is the rule recommended for deciding references on published documents. Users other than publishers may prefer other citation orders. Further, more detailed guidance on classification, unlikely to be required by most users, is given in Appendix 2.

Application 2: General information

2.12 How to file

The term 'filing' means here not only arranging individual documents on library shelves or elsewhere, but also arranging part documents such as data sheets or drawings in sets and arranging the chapters, paragraphs, etc of books. Codes make convenient labels for filing purposes because they have an ordinal value, ie 1 files before 2, A before B. In addition, use of CI/SfB codes for filing brings similar subjects together.

Before using the codes for arranging items, the arranger must decide the most suitable filing order for the categories which concern him. One filing order which is widely used is the order of the tables 0 to 4. This divides the items to be filed into 5 main groups, corresponding to Tables 0, 1, 2, 3 and 4. Parts and accessories follow the things to which they belong. For each item, the symbol furthest to the left indicates the group in which it should be filed eg:

27(21)Fg2   File in Table 0 at 27
  (21)Fg2   File in Table 1 at (21)

The remaining symbols define the precise filing position for the item. An item coded (21)Fg will file between items coded (21)Ff and (21)Fh.

Figure 25
Descending order of items in a schedule is equivalent to left to right order of documents on shelves.

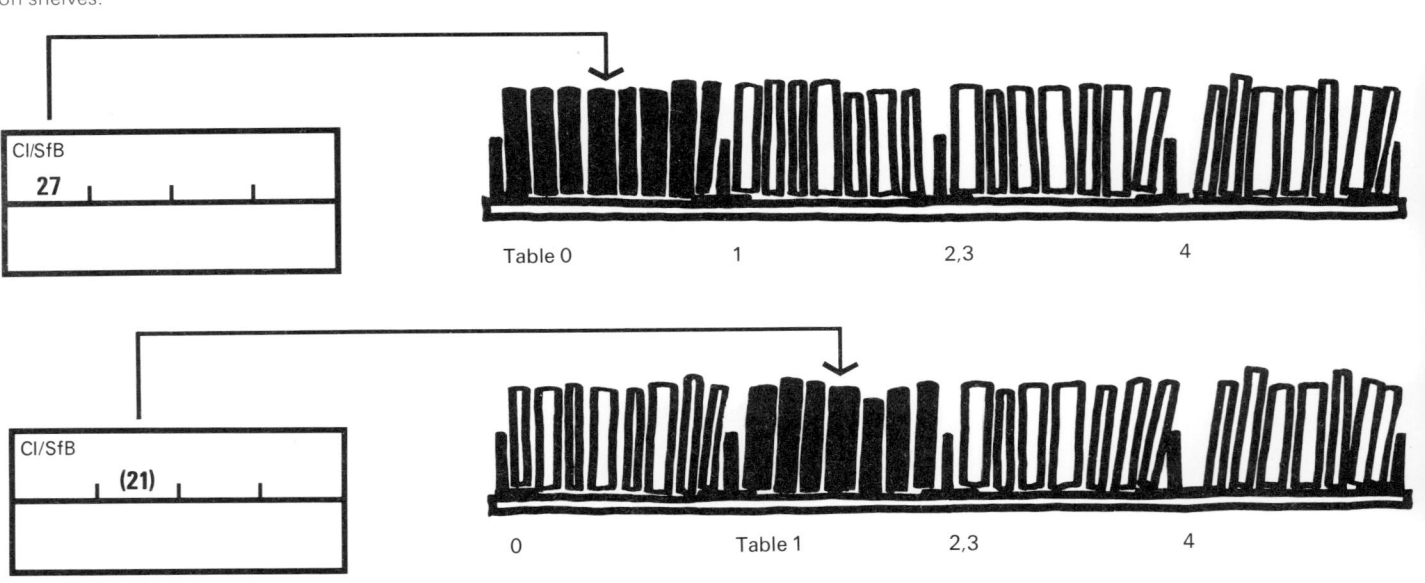

Freedom to choose
Any appropriate filing order can be employed, but where there are no special requirements, the order of the tables 0, 1, 2, 3, 4 is usually used. Where the primary filing symbols chosen are not those which occur first in the published reference, the symbols can be written out in the bottom half of the classification box in the required different order, eg:

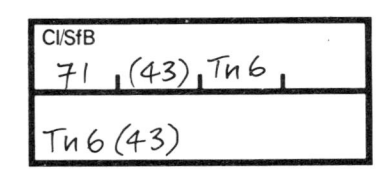

Application 2: General information

Headings without codes
CI/SfB provides a list of subject headings in a standard order, and these headings can be used without codes. Table 4 is already used in this way, to provide a standard arrangement for the content of product data sheets. Authors can use CI/SfB categories either as main headings or sub-headings eg:

Economics
External walls
Internal walls
Floors
Stairs

Also, CI/SfB headings can be used to name files which can then be kept in alphabetical order. Detailed advice on filing, unlikely to be needed by most users, is given in Appendix 2.

2.13 How to find

1. Find out what citation and filing rules have been used for the library (or document) being consulted. These will show, for example, whether information on floor finishes in schools is more likely to be filed with schools in Table 0, or floor finishes in Table 1. Usually the citation and filing orders will be in the order of the tables, with Table 1 following Table 0, and so on.
2. Decide the classification and code for items likely to be useful by going through the same steps as the classifier. If citation and filing order in the order of the tables has been used, the search code for information on floor finishes in schools is: 71 (43).
3. Look for suitable information at this code position.
4. If nothing is found, broaden the search by scanning other items in class 7 (not only those coded 71), including those subcoded with the broad code for finishes, ie: (4–), not only those subcoded (43).
5. If there is no information in class 7 on (43), look for information at class (43), – or as a final resort (4–) – to see if there is something there which might be useful for schools.

Indexes are valuable time savers for users who want specific information (rather than simply to browse) and often lead quickly to information which might not otherwise have been found. They save hours spent hunting through trade literature in library files, drawings in filing cabinets and chapters in books. The simplest form of index is a list of contents. The contents page in a book or a drawings list presents a simple record of the individual items or groups of items in the order in which they are filed. A contents list should be provided for every document or set of documents.

The main index in this manual is a general alphabetical subject index to Tables 0, 1, 2, 3, 4. This index may also be used as a guide to information in libraries and other collections arranged by CI/SfB, but it is not designed for this purpose. It cannot be as precise as the index compiled specially for a particular collection of information and its use will be misleading as it will include some terms that are not represented by items in the collection. It will equally omit many that are. Each individual document or set of documents should have its own alphabetical subject index. Indexes of trade names, manufacturers' names etc. may also be necessary. Two techniques of subject indexing; chain indexing and post-coordinate indexing, are described in detail in Appendix 3.

Application 2: General information

2.2 Published trade and technical literature

BS 4940 (Note 2.2) recommends that product information should be classified by CI/SfB **before publication,** and a central classification service is provided by the SfB Agency (Note 2.3) in order to achieve a reasonable standard of consistency in trade literature classification. General technical information can also be classified by the Agency.

2.21 Product or application classification

Most trade literature is not only about the product itself, but is also about the use or uses to which it can be put. Literature on bricks is usually implicitly or explicitly about their use in **brickwork,** literature on boilers may be about their use in **central heating systems,** literature on felt-backed pvc sheets may be about their use as **floor finishes,** literature on floor finishes about their use in **kitchens.**

Product (what it is)	**Application** (what it is for)
bricks	eg: for brickwork; or floor finishes
doors	eg: for external openings; or cupboards
boilers	eg: for central heating; or hot water supply
storage fittings	eg: for kitchens; or laboratories

People searching for products frequently look for information by product application, using their libraries as problem-solving tools (eg: what central heating methods are available?) but sometimes try to go straight to a type of product – (eg: what boilers are available?). CI/SfB has so far been able to classify by application, but not always by product. With this edition, product classification on a wider basis becomes possible if users want it. For the time being, the SfB Agency will continue to classify by application whenever possible.

2.22 Basic reference

The basic reference given by the Agency in the top right hand corner of literature is kept reasonably short, for use in the great majority of small libraries. The reference will usually be based mainly on Tables 1 and/or 2 and/or 3 of CI/SfB. It will follow the recommendations given below and in General information 2.1 'How to classify', and will be set out in the four-part classification box. The box is to a standard size 20 × 45 mm. It is shown twice this size below.

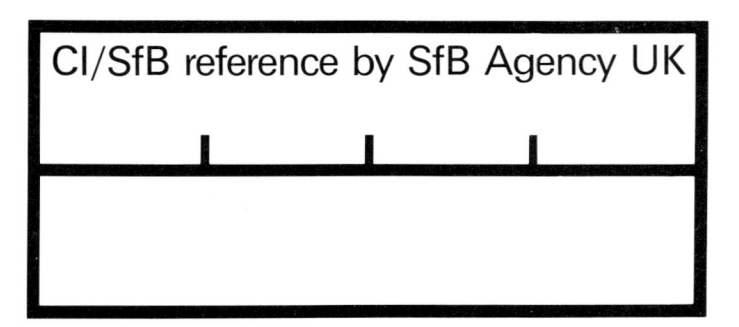

CI/SfB reference by SfB Agency UK

2.2 BS 4940:1973 'Recommendations for the presentation of technical information about products and services in the Construction Industry' obtainable from BSI Sales Branch, 101 Pentonville Road, London N1 9ND.

2.3 Full details of the service can be obtained from the SfB Agency UK, 66 Portland Place, London W1N 4AD telephone 01-637 8991 or at the RIBA, 01-580 5533.

Application 2: General information

The special indication 'Reference by SfB Agency UK' is printed in it **only** if the Agency itself has given the reference. The bottom half of the box may be used for Universal Decimal Classification (UDC) references (Note 2.4). UDC is a general classification, not a special construction industry classification, and surveys have shown that, of the two systems, CI/SfB is used to a far wider extent for office libraries. However, publications which are likely to have a longer life than trade literature and may find their way into large general collections, usefully bear an alternative UDC classification as well as a CI/SfB reference.

2.23 Coordination with specifications

Trade literature will often relate to specific divisions of the National Building Specification. In the examples of classification below, the first digit of the CI/SfB basic reference corresponds with the appropriate NBS main division code:

Subject of literature		Basic reference (by application)	NBS Division
Concrete for	**cast work**	Eq4	E
Clay bricks for	**brickwork**	Fg2	F
Steel mesh for	**brickwork**	Fh2 (see below)	F
Steel mesh for	**meshwork**	Jh2	J
Clay pipework for	**drainage**	(52)Ig2	5

The full code from Tables 2 and 3 for steel mesh reinforcement as an accessory product for brickwork would logically be F: Jh2, with F standing for brickwork (the application) and Jh2 for steel mesh (the accessory product). This is, however, a long and unwieldy code for library use, and the SfB Agency basic reference will instead use the short form recommended by the SfB Development Group of CIB. This involves attaching the initial digit of the application code direct to the last two digits of the product code, giving Fh2 in the case above.

This type of code will look unfamiliar to UK users, but it is international, is in scale with the filing needs of practitioners offices and has the merit that it permits consistent length coding from Tables 2 and 3 – a feature of particular value in computer applications. In every case above the Table 2 code indicates a type of work or construction, and the Table 3 code indicates the substance of the materials used in it. Clearly, Fh2 could refer to steel blockwork as readily as to steel mesh in brickwork, but experience suggests that this will not be a cause of real difficulty in practice.

2.24 Amplified reference

Trade and technical literature may carry a detailed classification based on the methods set out in Appendix 2, and printed vertically in the lower left hand margin, as indicated in BS 4940. An example of a detailed classification would be F: Jh2 from the case discussed above.

2.4 UDC references may be obtained from 'ABC (Abridged building classification for Architects, builders and civil engineers: a selection from the Universal Decimal Classification)' published 1955, with supplement 1965, by CIB (International Council for Building Research Studies and Documentation), Rotterdam and available from British Standards Institution (Overseas Sales), 101 Pentonville Road, London N1 9ND.

Application 2: General information

2.25 1968 references

References in accordance with the 1968 edition, where they differ from current CI/SfB references, should continue to be given on documents published during the year following publication of this edition. They should be given in small lettering in the bottom right hand corner of the front cover of literature, or in the bottom half of the classification box.

2.26 Series publications

The technical press and other organizations with series publications may wish to use CI/SfB in a flexible way to meet their users' specific needs. The SfB Agency should be consulted in such cases (see Appendix 2).

2.3 Office libraries

Many office libraries in the construction industry will have only a few documents on most subjects and will find that the basic references given in the top right hand corner of published trade and technical literature will adequately meet their needs. Figure 26 shows the growth of the Table 1 part of an office library, assuming filing in accordance with General information 2.1 on 'How to file'.

Most office libraries, when fully grown, will have individual files at main class level eg: 71 **School facilities**, and some for sublasses eg: 711 Nursery school facilities. The contents of such files can be organised systematically, where necessary, in a number of ways. For example:

1. Using the same principle as the CI/SfB Tables themselves, so that parts and accessories follow the things to which they belong. The file on 71 School facilities might be broken down with dividers into:
    712 Primary school and middle school buildings, spaces, parts (elements, products, materials), activities and requirements.
    713 Secondary school buildings, spaces, parts (elements, products, materials), activities and requirements
    to
    718 Other school buildings, spaces, parts (elements, products, materials), activities and requirements for schools generally.
    If an office file 71 has a lot of information relevant to only one subclass eg: 712 Primary schools, this can be given a file of its own. Information on other types of schools can file at 71 until further subdivision becomes necessary, or be kept in a file coded 713/718.
2. The contents can be subdivided alphabetically at any level by subject term, publisher or manufacturers' names.

Whatever method of arrangement is used, it is important to pursue a consistent policy throughout the library. Every office library, whatever its size, should consider whether or not to provide appropriate forms of alphabetical, subject or name indexes to its stock (see General information 2.1). Where they are provided, inevitably at some expense, technical staff should be asked to use them – and thus save valuable time every time they search for specific information – unless they know before hand exactly which document they want and where to find it. Librarians with large office libraries will find detailed advice on classification and filing in Appendix 2 and on indexing in Appendix 3.

Application 2: General information

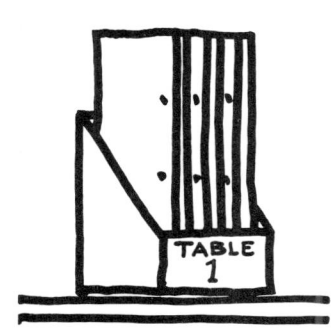

Stage 1: One file containing all information on elements

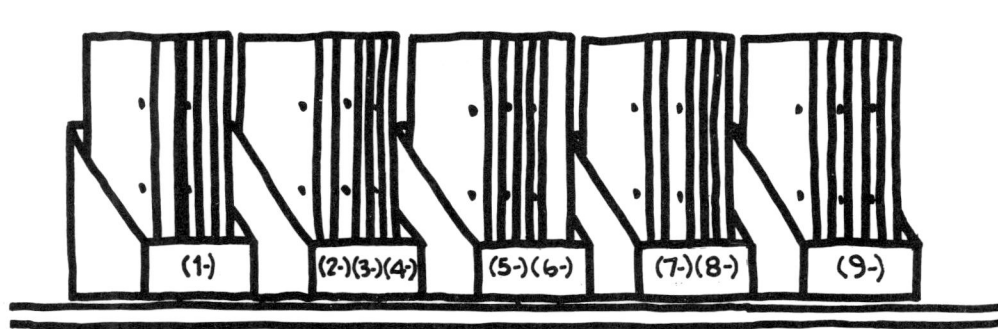

Stage 2 — Substructure — Building fabric — Services — Fittings — External works

(1-) (2-)(3-)(4-) (5-)(6-) (7-)(8-) (9-)

Sub-structure — Primary elements — Secondary elements — Finishes

(1-) (11) (13) (16) (17) (2-) (21) (22) (23) (24) (27) (28) (3-) (31) (32) (33) (34) (35) (37) (4-) (41) (42) (43) (44) (45) (47)

Services — Installations — Fixtures — Loose equipment — External works

(5-) (51) (52) (53) (54) (55) (56) (57) (6-) (62) (63) (64) (66) (7-) (71) (72) (73) (74) (75) (76) (8-) (81) (82) (83) (84) (85) (86) (9-) (90)

Stage 3

Figure 26
As the library expands, CI/SfB provides the extra detail required, as exampled by the RIBA Office Library Service. As a general guideline more detailed codes should be used only when several documents bear an identical code.

Appendices

1 Common element parts

Common element parts are an addition to CI/SfB identified by the SfB Agency partly as a result of the work carried out on the Construction Industry Thesaurus (CIT). The list provides a detailed subdivision for individual parts which may occur in several elements or groups of elements, eg columns, beams, pumps, and therefore cannot always be given the same element code. The breakdown provided will be too detailed for non-specialised office libraries. The codes are **not** therefore for use in the classification box on preclassified published documents. The main use of the list is likely to be as a standard order rather than as a coding system (as the CIB Master Lists), and there are potential uses in the arrangement of engineering libraries and for bibliographical and other specialised purposes within the Construction Industry.

The list includes some categories also included in Tables 2 and 3. Examples are (:36) Pipes as guiding devices, duplicating I in table 2 – pipes as forms of products and assembled work; and (:94) which parallels t6 in table 3. Users will need to decide which list best meets their particular needs, and adapt or omit the other accordingly.

Any individual element in Table 1 may have:

- (:1) Processing parts
- (:2) Containing parts
- (:3) Moving, guiding parts
- (:4) Controlling, measuring, recording, detecting, indicating parts
- (:5) Dividing, protecting parts
- (:6) Access parts
- (:7) Structural parts
- (:8) Covering, infilling parts
- (:9) Jointing, fastening, securing, operating parts, finishings, accessories

For example:

(56) Space heating services
- (56:1) Processing parts eg: boilers, calorifiers, radiators
- (56:2) Containing parts eg: tanks
- (56:3) Guiding parts eg: pipes, flues
- (56:4) Controlling parts eg: thermostats
- (56:5) Dividing parts eg: lagging

Intermediate ad hoc sequential or systems-related categories such as P(plant), D(distribution), F(fittings) can be used to separate radiators from boilers and bring flues and boilers together:

(56) Space heating services
- (56P)
  - 1 Processing parts eg: boilers
  - 3 Guiding parts eg: flues
  - 5 Dividing parts eg: lagging
- (56D)
  - 1 Processing parts eg: calorifiers
  - 3 Guiding parts eg: pipes
  - 5 Dividing parts eg: lagging
- (56F)
  - 1 Processing parts eg: radiators
  - 4 Controlling parts eg: thermostats

In this case the code for lagging to the distribution system for a central heating service would be (56D:5).

Appendix 1: Common element parts

(:1) Processing Devices
Relevant to two or more elements
If necessary subdivide eg:

(:11) **Destroying, disposal, reducing devices eg:** crushers, grinders; incinerators; chemical disposal devices; shredders

(:12) **Producing, supplying devices**
  1 Humidifiers, dehumidifiers
  3 Heat producing, supplying devices
    1 Heat producers and emitters eg: panelled, strip etc; contact, convectors, radiators; fires, stoves, furnaces, ovens; storage, instantaneous; heat pumps; electric
    2 Heat producers eg: fluid heaters, boilers, immersion heaters, calorifiers, steam heaters; air heaters
    3 Heat emitters eg: refrigerators; coolers
    7 Heat exchangers
    8 Other heat producing, supplying devices
  4 Lighting sources (luminaires)
  5 Signal producers and sensors
    1 Signal producers eg: bells, buzzers, chimes, loudspeakers, sirens
    2 Sensors eg: microphones, antennae aerials
  6 Electric energy producing devices eg: voltaic cells, generators, dynamos
  7 Mechanical energy producing devices eg: motors
  8 Other producing, supplying devices eg: reproducing devices

(:13) **Converting, stabilising devices**
  1 Compressors
  2 Condensers, evaporators
  5 Amplifiers
  6 Transformers, commutators, converters, inductors; receivers
  7 Transmissions, gears (other converting, stabilising devices eg: counter balancing devices)

(:14) **Mixers, separators**
  1 Mixers
  2 Separators eg: screens, filters, traps, sorters
  8 Other mixers, separators

(:15) **Cleaning devices eg:** washers, sterilisers

(:16) **Driers**

(:18) **Other processing devices**

(:19) **Parts, accessories etc special to processing devices** may be included here if described separately from specific types above

(:2) Containing devices
Relevant to two or more elements
If necessary subdivide eg:

(:21) **Waste containers eg:** liquid waste containers; refuse containers

(:22) **Solid/liquid separation containers eg:** sedimentation tanks, sediment pans

(:23) **Fluid containers eg:** cisterns, sumps, vats, tanks; sinks, basins, baths, troughs; air vessels

(:24) **Enclosing/partially enclosing containers eg:** cabinets, cupboards; chests; boxes, drawers; bins, bunkers; cannisters; bags, sacks; buckets, baskets

(:25) **Non-enclosing containers eg:** shelves, racks, trays

(:28) **Other containing devices**

(:29) **Parts, accessories etc special to containing devices** may be included here if described separately from specific types above

(:3) Moving, guiding devices etc
Relevant to two or more elements
If necessary subdivide eg:

(:31) **Distribution devices eg:** sprinklers, drenchers, sprays; diffusers, distribution boards

(:32) **Transmitting devices eg:** transceivers, telephone sets, handsets; transmitters; exchanges

(:33) **Impelling devices eg:** fans, pumps; smoke, fumes, air extract devices

(:34) **Conveying devices eg:** lifting devices, pulley blocks

(:35) **Moving devices eg:** reflectors, refractors; pivots

(:36) **Guiding devices**
  1 Ducts eg: conduits, flues, chutes, pipes, channels
    See also Table 2 Class I
  2 Conductors and resistors
    1 Conductors eg: circuits; electrodes; coils; solenoids; windings; contacts; connectors; lightning conductors
    2 Resistors
  3 Guide rails, tracks, handrails
  4 Steps
  5 Inlet, outlet devices eg: manifolds; inlets, including gullies, hoppers; outlets, vents eg: single point, multipoint; burners; exhausts; hydrants; outfalls; overflows; nozzles; spouts; socket outlets, terminals
  8 Other guiding devices

(:38) **Other types**

(:39) **Parts, accessories etc special to moving, guiding devices** may be included here if described separately from specific types above

Appendix 1: Common element parts

(:4) Controlling, measuring, recording, detecting indicating devices
Relevant to two or more elements
If necessary subdivide eg:
- (:41) Controlling devices
    1. Fluid flow control devices eg:
       valves eg:
       mixing valves, taps etc;
       weirs; dampers
    2. Humidistats
    3. Thermostats
    6. Electrical control devices eg:
       relays; protection gear eg:
       earthing, fuses; switchgear eg:
       switches, circuit breakers,
       programmers; overload
       protection devices
- (:42) Measuring devices
    1. Meters eg:
       potential energy meters eg:
       strain gauges; fluid meters eg:
       gas meters, volumetric, velocity
       and pressure meters; moisture
       meters; thermometers;
       electrical meters eg: integrating
       meters, ammeters, voltmeters,
       ohmeters
    2. Clocks
- (:43) Recording devices, recorders eg:
    loggers
- (:44) Detecting devices, detectors eg:
    fire, combustion detectors;
    temperature change detectors;
    intruder detectors
- (:45) Indicating devices, indicators eg:
    pointer type indicators,
    cyclometer type indicators
- (:48) Other controlling, measuring, recording, detecting, indicating devices
- (:49) Parts, accessories etc specific to controlling, measuring, recording, detecting, indicating devices may be included here if described separately from specific types above

(:5) Dividers, barriers, protecting devices
Relevant to two or more elements
If necessary subdivide eg:
- (:51) Vertical barriers, dividers, protecting devices eg:
    walling, fencing, hoarding,
    balustrading, bulkheads
- (:52) Horizontal barriers, dividers, protecting devices eg:
    decking
- (:53) Barriers, dividers, protecting devices according to specific hazards
    1. Wind barriers
    2. Moisture barriers, protection devices eg:
       damp-proof courses, tanking, water bars, creasings
    3. Fire barriers
    4. Thermal barriers eg:
       lagging, thermal insulation
    5. Sound barriers eg:
       reflectors, absorbers, sound insulation
    7. Crash barriers
    8. Other barriers, dividers, protecting devices according to specific hazards
- (:58) Other barriers, dividers, protecting devices eg:
    baffles; parting slips; guards, gratings
- (:59) Parts, accessories to barriers, dividers, protecting devices may be included here if described separately from specific types above

(:6) Accesses
Relevant to two or more elements
If necessary subdivide eg:
- (:61) Windows (devices of any size for admission of light and/or vision)
    1. Fixed
    2. Side-, top-, bottom-hung (ie hinged)
    3. Pivoting, projected, louvred
    4. Horizontal, folding, sliding
    5. Vertical, sliding
    7. Inward, outward-opening
    8. Other windows
    9. Parts of windows eg:
       inner windows (for double glazing); lights, sashes, casements;
       window frames, frames, boards, surrounds, furniture, glazing
- (:62) Doorsets, doors (devices of any size for admission of objects)
    2. Hung (ie hinged)
    3. Pivoting, including louvres, projected, revolving
    4. Sliding, folding, telescoping
    5. Vertical, sliding, folding, roller shutters
    6. Overhead opening, up and over
    7. Inward, outward-opening
    8. Other doorsets, doors eg:
       casement; flexible; fire-check; flush, sectional, panelled, matchboarded, braced, ledged, framed
    9. Parts of doorsets, doors eg:
       door frames, jambs, furniture
- (:63) Gates, grilles, hatches
- (:64) Limited access parts eg:
    manholes, rodding eyes, soot doors
- (:66) Vertical access parts eg:
    steps, stepirons, hooks, ladders
- (:68) Other accesses, openings
- (:69) Parts, accessories special to accesses may be included here if described separately from specific types above eg:
    opening leaves, frames, jambs, hardware (ironmongery)

Appendix 1: Common element parts

(:7) Structural members
Relevant to two or more elements
If necessary subdivide eg:
Structural members subject to axial and lateral forces (:71)/(:73)
- (:71) **Frames**
  1. Plane frames eg: portal frames, trusses
  4. Space frames
  5. Squared frames
  6. Moulded frames
  8. Other frames
  9. Parts of frames eg: stiles, jambs, studs, mullions, muntins, rails, noggings, heads, transoms, sills
- (:72) **Arches**
  1. Straight arches, gauged arches
  2. Curved arches
  3. Queen Anne arches (Venetian arches)
  4. Arches according to number of centres eg: single centre arches including segmental, semicircular (barrel)
  5. Parabolic arches; relieving arches
  8. Other arches
  9. Parts of arches eg: intrados, extrados, springing lines; crowns, haunches; voussoirs; ribs
- (:73) **Buttresses; abutments; other members subject to axial and lateral stress**
- (:74) **Lateral force resistant members**
  1. Beams (flexural members) eg: purlins; lintels; joists; sloping beams, rafters; box, plate, vierendeel beams; girders (mainbeams), secondary, tertiary beams
  2. Cantilevers (flexural members)
  3. Shear force resistant members
- (:75) **Axial force resistant members**
  1. Struts (compression members) eg: columns, pillars, stanchions; pedestals
  2. Ties (tension members) eg: wall ties, collars
- (:76) **Supports (vertically loaded structural members) eg:** bearers, seatings, mountings, plinths, bases, brackets, corbels, bearing plates, hangers
- (:78) **Other structural members eg:** bracings; stays, props; stiffeners, diaphragms; piers, posts; rakers
- (:79) **Parts, accessories etc, special to structural members, may be included here if described separately from specific types above eg:** reinforcement

(:8) Coverings, infillings
Relevant to two or more elements
If necessary subdivide eg:
- (:81) **Outside coverings eg:** claddings, skins, sidings
- (:82) **Linings (inside coverings)**
- (:83) **Caps (top coverings) eg:** lids, copings, sills
- (:84) **Overlays (overcoverings) eg:** cornices, carpets; mats, veneers; wearing courses; joint overcoverings eg: bosses, flashings, architraves, joint boxes, coves
- (:85) **Underlays (undercoverings) eg:** base courses; joint undercoverings eg: saddle pieces, soakers
- (:86) **Infillings eg:** spandrels, bricknogging, glazing
- (:87) **Surrounds, edge coverings eg:** nosings, verges, fascias
- (:88) **Other coverings, infillings eg:** sheaths, cover plates, cover fillets, shutters, drapes (shrouds), hoods
- (:89) **Parts, accessories special to coverings, infillings, may be included here if described separately from specific types above**

(:9) Jointing, fastening, securing, operating devices, finishings, accessories
Relevant to two or more elements
If necessary subdivide eg:
- (:92) **Jointing devices, joints**
  See also Table 2 class Z
- (:94) **Fasteners, fixings**
  See also Table 3 class t6
  Anchoring devices eg: anchors; grounds, inserts, fixing slips, plugs
  Attachment devices eg: clips, clamps, cleats, connectors, couplings, hinges
  Fixing devices eg: rivets, nails, nuts, bolts, screws, studs
- (:95) **Securing devices**
  See also Table 3 class t7
  1. Locks eg: padlock, hasp and staples; rim, recessed, mortice locks; single-, double-sided locks; dead locks; tumbler, lever operated, spring operated, longthrow locks
  2. Catches eg: ball, elbow, gate, gripper, magnetic, roller, hooks and eyes, pegstay, cockspur, button, counter flap catches
  3. Latches eg: thumb latches
  4. Bolts eg: sliding, panic, cremorne bolts; espagnolette bolts; shoots
  8. Other securing devices
- (:96) **Operating devices**
  1. Operating devices by function eg: opening, closing
  2. Remote controlling devices
  3. Manual operating devices eg: handles, buttons, discs, knobs, switches, capstans, rings, wheels, levers (keys), cords; muscle potential, hand, foot, knee operating devices
  4. Automatic operating devices
  5. Pressure operating devices eg: hydraulic, pneumatic operating devices, heat operating devices, light operating devices, sound operating devices, electrical operating devices, mechanical operating devices
  8. Other operating devices
- (:98) **Finishings, ornaments, trims, mouldings**
- (:99) **Accessories**

2 Classification and filing

Based on the section by J. Mills FLA in 'The organisation of information in the construction industry' (SfB Agency UK Development Paper No. 3) RIBA 1968 (out of print).

2.1 Basic principles of information retrieval

A collection of documents or items, however modest, is a store of information. **Information retrieval** describes the operation of locating the items within the store which are relevant to a request for information. When searching, it is only practical to examine a small set or class of documents, or items. The question is: which shall it be? Indexing is the process of indicating or pointing out the whereabouts of a particular type of information. This implies the prior recognition of classes of information so that the class likely to be relevant to a request can be distinguished from others. The recognition of classes and of relationships between them is the fundamental role of classification in information retrieval.

2.11 Indexing

Indexing demands initially two basic steps:
1. The indexer decides what concepts make up the subject of the document. This is usually done by extracting from the title, summary and text, the keywords which best convey its subject, eg: domestic – stairways.
2. These terms are translated into a controlled index language. Domestic – stairways, for example, may need to be translated into houses – stairs.

At this point there are several ways of proceeding. In most libraries, the physical arrangement of the documents themselves is used to form the main index. The documents can be grouped according to their subjects, so that those dealing with a particular subject will as far as possible be found together. This involves the classification of the document as an entity, to be assigned to one particular class rather than another. Stairs in houses for example would be treated as a single class to be located with the class houses (in general) or with the class stairs (in general), but not with both. As an index this shelf classification has certain limitations. It is therefore usually supplemented by a catalogue of some kind in which the entries standing for the documents are often arranged in the same order as the documents themselves. An A/Z index of the terms used is included in such a catalogue.

In recent years a new form of indexing, co-ordinate indexing, has developed in which the keywords applying to each document are recorded separately and terms are brought together only if a question is received which requires this co-ordination. Each method of indexing has relative advantages and disadvantages and these are described below and in Appendix 3.

2.12 Searching

When a document is indexed by subject, it is metaphorically labelled with tags announcing that it might be relevant to questions on x, y, and z, being the concepts it mainly deals with. For example, a report on condensation in roof sheeting is labelled as containing information on the process condensation, on the construction form sheet, and on the part of the building roof finish. Each enquiry must then be analysed and labelled in the same way, so that the store (Note 2.1) can be searched by means of its index and a match sought between the tags describing the question and those describing the documents. If both sets of tags have been couched in

---

2.1 The words 'store' or 'collection' are used where possible in preference to 'library'. The explanation assumes a document in a collection, but the principles apply equally to a paragraph or line within a document, or an item of information in any other form which may be organized by CI/SfB.

Appendix 2: Classification and filing

the same terms the chances of a match are much better than they would be if synonymous but different terms were used. A request for information on stairs for houses would be 'programmed' by CI/SfB and sought under the class houses, sub-class stairs. If a document on this subject was held it would have been similarly programmed and placed (Note 2.2) where expected, and not under stairs, sub-class houses.

It often happens that the question does not match the documents exactly. For example, a question on condensation in roof finishes in general might be put to a collection in which the nearest matching class (the most appropriate document in the collection) is on condensation in roofing sheets. The indexing language must help the searcher to adjust his search to what the collection actually has, to assist him to widen it if he is finding too little, and narrow it if he is finding too much. The search for information on condensation in roof finishes might be extended by examining the different kinds of roof finish.

This would include classes designated by their material or their form, eg: sheets. Each of these classes could be examined to see if it had a sub-class condensation, at which point a document labelled: roof finishes – sheets – condensation should be retrieved. If this failed, documents on roof-finishes-according-to-material could also be examined to see if they included information on condensation. Since a roof finish is a building element, other likely members of this class (eg: walls, finishes) could be examined to see if they had sub-classes on condensation or, less specifically, on water and moisture properties. On the other hand the concept of condensation itself could be pursued; all classes relevant to vapour and liquid could be examined to see if the documents in them referred to condensation.

Searching in this thorough, methodical way is greatly assisted by a simple alphabetical (A/Z) subject index. Some of the kinds of references such an index might provide are:

condensation	see (L2)
condensation: sheets, overlap: roof finishes	see (47) N(L2)
condensation: tiles: wall finishes	see (42) S(L2)

2.13 Indexes

An index indicates documents and classes most likely to answer a given question. Usually the term is used with the narrow meaning of a list of references which omits all details of author, title, etc. Two other forms of index are the physical arrangement of the documents themselves on the shelf; and the catalogue.

Shelf arrangement as an index: Scatter
In a public library arranged by the Dewey Decimal Classification, books on building construction will be found in class 69; general works are marked 690 and are followed immediately by sub-classes like building materials at 691, timber at 691.1, stone at 691.2 and so on. In this way, documents on closely related classes are found near to each other.

2.2 Assuming the use of CI/SfB for pre-classification in this case, rather than coordinating indexing.

Appendix 2: Classification and filing

```
(47)N(L2)
Condensation in sheeted roofs, by A.W. Smith.
BRS 1958.
(National building studies - Research papers - 23).
(47)N(L2)
```

1. Subject entry (the heading is the CI/SfB class number for Roof finishes – overlap sheets – condensation)

```
Smith A.W.
Condensation in sheeted roofs, by A.W. Smith
BRS 1958.
(National building studies - Research papers - 23).
(47)N(L2)
```

2. Author entry

```
Condensation in sheeted roofs.
Condensation in sheeted roofs, by A.W. Smith
BRS 1958
(National building studies - Research papers - 23).
(47)N(L2)
```

3. Title entry

```
National building studies -
Research papers - 23.
Condensation in sheeted roofs, by A.W. Smith
BRS 1958
(National building studies - Research papers - 23)
(47)N(L2)
```

4. Series title entry

Shelf arrangement is an important index in its own right. The quality of the arrangement (how well it keeps together material which is closely related) determines how easily the adjustment of search described in Appendix 2.12 can be made. But shelf arrangement has two serious weaknesses as an index:

1. Related items are scattered for largely physical reasons; drawings, for example, cannot easily be shelved with books and may have to filed separately in boxes. Technical information may be shelved separately from trade literature on the same subject, and some of the material may be on loan, so missing from the shelf.
2. Related items are scattered because of the impossibility of classifying by more than one characteristic at a time. If a document on condensation in roof finishes is shelved with others about parts of buildings, (like roof finishes) it cannot also go with documents about water vapour problems and moisture, unless two or more copies are kept. A technical way of putting this is that documents cannot be classified by 'part' and 'process' at the same time. It is only possible to divide by one characteristic first (by part, say, to keep together all documents on a particular part) and then by another (here, process, to keep together, under each part, the documents on a particular process relating to that part).

This dilemma is quite unavoidable. Whatever arrangement is adopted, many closely related documents will be separated. In the local library, for example, it may be necessary to look under a dozen or more classes to find all the documents on Georgian architecture:
under Architecture – GB – Georgian;
under Architecture – Churches – GB – Georgian;
under Architecture – GB – London – Georgian;
under Architecture – Houses – GB – Georgian;
under Architecture – Houses – Country houses – GB – Georgian, and so on.

Classes which are related yet are scattered in this way, are called 'distributed relatives'.

The catalogue as an index
The essence of a catalogue is that it uses small substitutes for the documents themselves, and by multiplying these, gets many different arrangements, thus overcoming the weakness of shelf arrangement. An example is the card catalogue, in which a card carries a brief description of a document (its author, title, publisher, date etc). Several copies may be made and each copy given a different heading. Each card (entry) includes the class number showing where the item is located on the shelves. Examples of this technique are shown on the left.

Using these substitutes, it is possible to make a representation of the collection arranged in different ways – eg: by subject in classified order, by author, by title, by series. The subject catalogue must be kept separate, since it is arranged according to the classification system, using the class numbers. But the other three catalogues could be merged to make one continuous alphabetical sequence, since their headings are all names, for which A/Z is the only feasible arrangement.

Appendix 2: Classification and filing

Catalogues have three major advantages over shelf arrangement as an index to resources:
1. They are comprehensive: there is no absence of material because it is on loan, or in a 'reference only' section, or oversize.
2. They allow 'multiple-access'. Readers may approach the collection with a request for items by a particular author; in a particular form (eg: directories); with a particular title; on a particular subject, and so on. Moreover, in the subject catalogue it is relatively easy to have more than one entry per document. A document on condensation in roof finishes, for example, could have one entry under roof finishes and another under condensation. Or, subject entries could be arranged alphabetically in a second file to complemennt the classified arrangement in the first.
3. They need not be restricted to one physical collection, but can indicate material housed elsewhere, or published in a given period or in a given subject area (eg: British Technology Index; RIBA Product Data; Index Medicus). Published indexes such as these can be cut up, pasted on cards and merged with the catalogue to a particular collection. In other words, they provide bibliographies, abstracting and indexing services and so on.

In smaller offices it is of course, impossible to provide all these additional catalogue-indexes. But once a collection has reached a certain size the provision of some sort of catalogue-index to supplement the shelf arrangement becomes essential if the collection is to be fully exploited.

2.2 Basic principles of classification
(designing a classification)

A class is a set of documents or concepts dealing with the same subject. The term is relative; a class may be a sub-class of a containing one, and itself contain sub-classes. Seven major features characterize a classification. These can be considered in the order in which they enter into the process of making a classification of any special subject field, since this is almost unchangeable, each operation depending on the prior performance of the preceding ones.

2.21 Facets (categories)

The field is first divided into broad categories. For example, if documents on the construction industries are divided by the principle **type of building**, classes such as house, hospital, high rise buildings, occur as distinct from elements (eg: roof, wall, lighting system, heating system); or **constructions** (eg brickwork); or **materials** (eg: stone, timber, metal), each of which have a different relationship to their containing class 'construction industries'. The total classes reflecting division by one broad principle is called a facet.

2.22 Sub-facets

Each broad facet is divided into precise sub-facets, eg: parts which are specifically installations or services, such as heating system, are sorted out from those which are to do with the fabric, eg: roof.

2.23 Citation order

'Citation order' refers to the order in which the elementary concepts within a single compound class are cited or to put it another way, the rule for deciding how to build up a compound class number for a particular

Appendix 2: Classification and filing

document. Most documents on the construction industries deal with compound subjects (which reflect more than one facet), such as kitchens in hospitals. This immediately poses the problem of which class number such a subject gets first – the one for kitchen or the one for hospital?

It has already been said that the central problem of shelf order and the conventional subject catalogue order is scatter. This problem is reduced by observing a consistent citation order. If a rule is made that parts of buildings are subordinate to types of buildings, the first example will automatically be put under hospitals – kitchens, not kitchens – hospitals. The type facet is cited before the part facet when describing a compound subject.

If a consistent citation order is observed, the locating of compound classes becomes predictable. So a modern classification system usually rules, or advises, that a citation order is to be followed, covering every facet and class in the scheme, making it clear where a document should go, whatever combination of notions it reflects.

2.24 Order of classes within sub-facets

At this lowest level of analysis, classes are equal or co-ordinate (mutually exclusive), since none is subordinate to another. When their filing order on the shelf or in a classified catalogue is considered, some helpful principle of arrangement should be found. For example, in the place facet of most classifications, a geographical order is followed so that places geographically near to each other file near to each other. In UDC, this is approximately west to east, beginning with Europe. Similarly, periods of history are arranged so that later periods follow earlier ones.

2.25 Filing order (shelf order, schedule order)

Filing order is the overall sequence of classes, single and compound, which determines the placing of already-coded items in catalogues or documents on shelves. The main determinant of course is citation order, which determines what is collected and what is scattered. The main problem is the filing order of the facets, eg do classes in the building types facet file before or after documents in the elements facet? The other problem is of order **within** subfacets, dealt with in Appendix 2.24. In the following example, the order of files given by the first part of the class number, 41 – hospitals; (23) – floors; (P2) – sound insulation; is exactly the same as the order in which these classes are given in the tables.

41	Hospitals	Building type
41(23)	Floors in hospitals	Building type ÷ element
41(23)(P2)	Sound insulation in floors in hospitals	Building type ÷ element ÷ requirement
41(P2)	Sound insulation in hospitals	Building type ÷ requirement
(23)	Floors	Element
(23)(P2)	Sound insulation in floors	Element ÷ requirement
(P2)	Sound insulation	Requirement

This order of classes downwards in a book form, classified bibliography, or catalogue, becomes an order of left-to-right for documents on shelves.

Figure 27

Appendix 2: Classification and filing

2.26 Notation

If the five features of classification described so far are observed, the result is a detailed, comprehensive and predictable sequence of classes. New subjects can be accommodated in their logical positions when necessary. So that the relative position of any given class can be seen quickly, every class likely to require separate identification by numbers of users needs to be represented by a symbol with a recognized value, as 1 precedes 2, or A precedes B. If this is done, so that every class has its class-number, the position of each document or item is quickly apparent. Such a system is called a notation.

Many users of library classifications think that notation, besides showing the relative position of each class, should also perform another function — that of expressing the hierarchy or showing how some classes are subordinated to, or are equal to, others. For example:

8	residential buildings
81	housing
811	single storey
8111	detached
8112	semi-detached (etc)
812	two storeys
85	communal residential
852	hotels

These class numbers indicate that housing precedes hotels, but they also seem to indicate that house is a kind of residence by making its class number 81 a decimal division of the number for residence 8; also, that detached 8111 and semi-detached 8112 are equal sub-classes of the class single-storey housing 811, and that housing 81 and communal residential 85 are equal sub-classes of the class residence 8. This is very useful; a user scanning a shelf can quickly move to a more general class in order to retrieve more information. Also, it has proved very useful in machine coding.

It is, however, very difficult to maintain the principle consistently. There are two main reasons for this:
1. There will be occasions when more than ten equal sub-classes occur, and there are only ten numbers, 0/9, available. In such cases, a distinct number is usually given to the major sub-classes and others are lumped together under shared numbers.
2. It may result in very long class numbers; whereas if notation is not expressive, very brief numbers can be used, for example:
   (E)   Composition (properties)
   (G)   Appearance
   (G4) Texture eg: flatness, smoothness
   If the notations were to be fully expressive the number for textural properties would be something like (EAA) — three characters instead of two. Because of this, most modern classifications do not attempt to give expressive notations except where it is easily done.

CI/SfB notation is faceted. Because the sets of notations for tables are distinctive they are immediately recognizable and can be combined in different citation orders if special needs seem to require this (but see Appendix 2.36). Although class numbers are sometimes long, they are easy to handle and remember, because of their mnemonic (memory-aiding)

Appendix 2: Classification and filing

quality. This is a result of the predictable repetition of the same symbol to represent the same concept wherever it occurs: external walls is always represented by (21) whether it appears in a compound with a particular building type, a form, material or property.

2.27 A/Z subject index

The relative position of every class is made clear by giving it a class number, but this is not much help unless the number for a given class is known to users. This is done by providing an alphabetical subject index for the classification, with the names of the classes (including synonyms) listed in A/Z order and their class numbers given alongside, for example:

Ventilation (L2)
Ventilation services (57)

An important further feature of the A/Z index is that it shows clearly if a term is a distributed relative; the fact that the concept 'ventilation' appears in several different locations is brought out in the above extract. It is important to distinguish the A/Z index to the printed schedules of a classification scheme from the A/Z index made by a library to its own collection. The former covers the whole vocabulary of the subject, and for the reasons of economy, can only show the location of relatively simple terms. It could not hope to show all possible compounds formed by the combination of terms from two or more facets, eg:

brick retaining walls (16)Fg
concrete block retaining walls (16)Ff
maintenance: concrete block retaining walls (16)Ff(W)

The possible number of such compounds runs into millions. So the A/Z index to printed schedules usually contains simple subjects only, and gives the class numbers for single terms such as blocks; bricks; maintenance; retaining walls, but not for compounds of these (eg: maintenance of retaining walls). The A/Z index to a particular collection can go a long way towards providing an index to all the subjects represented, both simple and compound. How this is done is described in Appendix 3.

2.3 Application of classification

As with all indexing systems, deciding on a Cl/SfB class number for a document always involves two separate steps:
1. Concept analysis: deciding what the document or item is about – what are the key concepts dealt with;
2. Translation: stating the concepts in the indexing-language – giving the document a Cl/SfB class number. The basic procedure is described in 'How to classify', (General information 2.1) and is treated only briefly here. This section deals in more detail with matters of interest to classification specialists or to librarians in large libraries.

2.31 Concept analysis

The object is to arrive at a 'summarisation' or concise description of the subject of the document as a whole, rather than an enumeration of the chapters, sections, paragraphs, etc which make it up (as the index at the back of a book tries to do). In most cases the title, if the document has one, is a reasonably reliable guide to this, although it will occasionally be misleading. But in any case, the classifier should always be prepared to go

Appendix 2: Classification and filing

beyond the title and quickly scan the text and other content. As soon as the classifier has decided the key terms which convey the precise subject, he can then proceed to the next step, which is its translation into a CI/SfB code.

2.32 Deciding the CI/SfB code (translation step)

Since each document must end up in only one place on the shelf or each item n only one place in the bibliography or other list, strict rules (Note 2.3) are needed for combining different parts of a compound class. For example:

A document on acoustic finishes for floors in flats is concept analysed in terms of CI/SfB categories as: building type? – flat; element? – floor finish; form of product? – none; material? – none; property? – sound proofing. So the summarization of the subject is:
flats – floor finishes – sound proofing
816   (43)   (P2)
and the CI/SfB code is 816(43)(P2).

A document on softwood sections for roofs of housing is concept analysed as:
housing – roofs – sections – softwood
81   (27)   H   i2
CI/SfB code 81(27)Hi2.

A document on corrosion resistance of non-ferrous metals is concept analysed as:
non-ferrous metals – corrosion resistance
h   (L4)
CI/SfB code h(L4).

2.33 Citation order in detail

Rules for citation order should cover every part of a classification scheme, within tables as well as between them, because a document may deal with two or more concepts from within the same table. For example, landscape for factories reflects two concepts from Table 0.

This raises two questions:
1. Which concept is regarded as the primary one (the first cited one)? Does the example above go under landscape 998 or under factories 282?
2. Can class numbers from the same table be added together in the same way as class numbers from different tables?

The answer to the first question is that class numbers within the same table can be added together (see Appendix 2.5). The answer to the second is rather more complicated.

Even if two or more different numbers from within one table are not added together (Note 2.4), there is a problem with citation order, ie which simple class number is to be chosen? If a class number which means specifically landscape and factories together is not given, but only a number meaning

---

2.3   If these rules (appendix 2.13, 2.23, 2.33, 2.37) seem over-complicated, it must be remembered that there are $3\frac{1}{2}$ million ways of arranging 10 documents in a collection. Similarly, if a subject were specific enough to demand 10 terms to describe it, there would be nearly $3\frac{1}{2}$ million different ways of combining them.

2.4   The Manual does not encourage the use of two or more numbers from the same Table on published literature, since this can lead to long numbers which will be too specific for most users.

Appendix 2: Classification and filing

either landscape or factories, a consistent rule is still needed to tell users which of the two it shall be.

The citation order between the tables reflects a basic order used in indexing literature. It is based on the theory that the object or purpose of studying a subject should give the primary (first cited) facet in the shape of the end product and that all the contributing factors which lead up to it are subordinate to it. In the construction industries, the end product is a physical environment or building complex or facility (eg space) of some type. A building consists of various parts of functional elements which make up the whole; these in turn consist of various constructions and products, which are themselves of certain forms and materials. All these things (buildings, parts, etc) possess certain properties and have to meet certain requirements and may display within themselves certain processes, eg: weathering, cracking, decay. In production, certain activities (operations) are undertaken (natural forces also play a part sometimes), and certain agents are used, such as people, organizations, plant, equipment, etc.

The whole series reflects the subordination of means to ends and of parts to the whole. Stated in general terms it reflects an order of citing the elements in a compound subject, as follows: (Anything) – its types – its parts – its materials – its properties – its processes – operations on it – agents of these operations. By and large, this sequence also governs the citation order of compounds within a table.

Types, parts, materials
A whole is cited before its parts, eg: wall opening – door – sliding – lining; houses – kitchen. The type or kind of a thing, sliding, is cited before a part, lining. A part may have its own types and parts, eg: in the class sanitary fittings – showers – automatically mixed, shower is a part, automatically mixed a type of that part.

Properties, processes
The distinction between processes and properties is usually clear, and tunnel – width, or plaster – hardness, are examples of properties, whereas processes are activities which go on within something, or between things, such as tunnel – heat transfer; or, plaster – cracking. But the significance of processes and properties is that both reflect features by which buildings, their parts and materials are judged to meet user requirements. Wood rot, for example, is a process going on in the wood; fire resistance of a timber beam could be regarded as a property in so far as it can be measured, but it implies a potential process in the timber under certain conditions. CI/SfB acknowledges this overlap between two theoretically distinct facets and includes both processes and properties in one sequence of factors in Table 4.

Operations
Operations are activities human beings (or occasionally, natural forces) perform on something, such as tunnel – excavation; or, roof – design. Just as there are types of a part (eg: door – sliding) as well as parts of a type (eg: door – sliding – lining) there are properties of processes and of operations. Examples are the speed of a plaster-setting process; the efficiency of a test; the cost of mechanization. Also, it is possible to have operations on kinds and parts of things.

Appendix 2: Classification and filing

In all these cases, properties, processes and operations are subordinated to the thing (concrete or abstract) to which they refer, for example:

cladding – durability	(thing – property);
plaster – setting – speed	(thing – process – property);
bricklaying – fatigue	(thing – operation – process);
wood – decay	(thing – process);

Operations on anything are cited after that thing, eg:

glass – handling	(thing – operation);
bricklaying – workstudy	(thing – operation – operation).

Agents of operations
The agent (tool, instrument etc) of an operation or process is cited after that operation or process, eg: glass – handling – equipment. The notion of agent is a ubiquitous one in indexing. The agent may be a definite piece of equipment, a material or a method, (eg: pipe-laying – computed load method). The end in each case is the efficient performance of an operation or process and agents are the means towards this end.

Closely related to the notion of agent is the maxim that when one thing affects or influences another, the thing affected is cited first, as with building materials – effects of heat – (by) solar radiation. The notion of condition is also analogous to an agent; for example, bricklaying – cold weather.

2.34 The 'and' problem

When indexing a document, the appearance of the word 'and' in the title demands that a careful distinction be drawn between two situations:

1. 'And' may be purely additive, showing that two separate themes occur in the document, eg: factory sites and foundations. In a subject catalogue such a document would get two entries eg: factories – foundations, and factories – sites. But this is no solution to the shelf location of the item in which the rule is to classify it under the theme most prominently dealt with, or under the appropriate summary class (see Appendix 2.37).

2. The 'and' may imply a relation between the two concepts, in which case a citation order must be decided between the two to subordinate one to the other. This is likely to be the effect or agent relation, eg: noise and privacy refer to the effect of noise on privacy and the citation order is privacy (effect of) – noise.

Observance of these basic indexing rules should ensure high consistency, but CI/SfB also includes an indication of the citation order within tables should more than one term from a table appear in the same document.

2.35 Citation order within tables and classes

The following is a summary of some of the major sub-facets between which an indexer may have to choose, whether this is simply to choose one and ignore the others, or to decide the order for adding one number to another.

Table 0
Classes 0 are cited before 1/8, eg: 06 or 06:28 land use – factories (note that a colon is used since the lack of it might result in confusion between 06 28 and 062 8). Classes 0/8 are cited before 9, eg: 81:93, housing – kitchens. Early classes are cited before later classes at every level, eg: 712:718 primary schools – classrooms (usually just 712 quoted).

Appendix 2: Classification and filing

Table 1
Early classes, eg: (2–) are cited before later classes, eg: (3–) and (4–). For example, (21)(31) curtain walling – window openings (usually just (21) quoted).

Tables 2 and 3
It can happen that two different constructions or products are referred to in a document in some definite relationship. In Tables 2 and 3, for example, metal-clay laths for plasterwork, aluminium edge trim for carpeting or bitumen damp-proof strip for brickwork, there is a relation between an application and a product, a primary function and a secondary function: the primary function of providing a plaster coating, and the secondary function of supporting it on laths; the primary function of providing carpeting, and the secondary function of trimming its edges; or the primary function of constructing brickwork, and the secondary function of making it damp-proof. According to recommended principles for citation, the secondary function, construction or product should normally be subordinated to (cited **after**) the primary one, and the methods used by the SfB Agency were described in General information 2.2.

The examples above would be coded from this viewpoint as follows:
metal-clay laths for plasterwork   Pr:Jh
aluminium edge trim for carpeting   T:Hh4
bitumen damp-proof strip for brickwork   Fg:Ln2

There are circumstances in which it may be preferable to omit steps in the theoretical analysis.

1. It is often the case that the secondary product is applicable (in the same relationship) to a number of other constructions or products and that many seekers after information on it would not necessarily associate it with the particular primary construction or product named, eg: the bitumen damp-proof strip could be used in construction situations other than brickwork, and the edge trim other than for carpeting.

2. If the notation is to be kept within reasonable bounds it may be desirable to cite only one construction form and material. If the primary construction form is cited first, the concept of most immediate interest goes unrecognized (the laths, the trim, the damp-proof strip). This is particularly the case where the library does not make an A/Z index to its material (which would be able to indicate the different locations of the secondary product).

3. The secondary product may be of immediate interest because it is very often this rather than the primary product which is being advertised or sold.

Appendix 2: Classification and filing

4. It may be unnecessary and unwieldy to distinguish between **closely** related concepts, eg: bricks for brickwork, in some applications of CI/SfB. In such circumstances, some concepts in the examples might be suppressed, as shown in brackets, giving codes as indicated:

Metal-clay laths (for plasterwork)	Jh
Aluminium edge trim (for carpeting)	Hh4
Bitumen damp-proof strip (for brickwork)	Ln2

Primary and secondary functions are also apparent in cases where Tables 1, 2 and 3 are used together, and here again it is possible to suppress concepts to give economical coding.

Asbestos sections for (casings) for (steel) building frames	(28)Xh2:(99:8)Hf6
use:	
Asbestos sections for building frames	(28)Hf6
Fire retardant (liquid) for coating of (steel) building frames	(28)Xh2:V:Yu4
use:	
Fire retardant coating of building frames	(28)Vu4
Asphalt coatings for floor finishes (for cast in-situ concrete floor beds)	(13)Eq4:(43)Ps4
use:	
Asphalt coatings for floor finishes	(43)Ps4

These examples are not intended as models. The description could be condensed in different ways. Where SfB Agency principles are not followed, it is important for any one collection of information that a decision is made in principle whether, and in what circumstances, concepts will be suppressed, or whether the fullest expression will always be used (which is less likely). As an example, a decision should be made as to the circumstances in whih a Table 2 code should be used to introduce a Table 3 code. Provided precedents are recorded – and the larger the collection the more essential it is that this should be a formal written record, usually on cards – the suppression of concepts should cause no problems.

Table 3
This contains two main sub-facets – materials by constitution (e/s), and materials by function (t/w). The latter includes some classes (eg: t6/t7 ironmongery) which are not strictly materials in the sense reflected by the rest of the table. When a topic combines both constitution and function (eg: copper/chrome preservatives) the function should be cited first – here, u3 preservatives, not h6 copper alloys.

Table 4
This table contains two major facets, activities and agents (A)/(D), requirements and properties (E)/(Y) – and a number of common facets – including place, time, form of presentation, all introduced by (Z) and using their UDC notations. (Z) is also used to introduce UDC notation for peripheral subjects.

Appendix 2: Classification and filing

The normal citation order, should two of these occur in the same topic, is: requirements – activities – place – time – form. For example, housing for old people in Scandinavia 81(U32)(Z(48)) if both numbers are used, otherwise 81(U32) omitting the reference to Scandinavia. This reflects the citation order of building type – requirement – place. The activities and agents facet has two features worth noting here:
1. Construction activities proper are cited before general activities such as administration, communication, training, eg: (D1)(A5) site protection, control procedures.
2. Although the general rule is to subordinate agents to operations it will be found that all construction equipment is gathered together at (B) and then divided by its operation.

Also, certain personnel (whose relation to the construction industry, is, of course, that of agents) are kept together at (Am) because their function is too general to warrant their subordination to a precise function: 'architect', 'quantity surveyor', are examples.

2.36 Alternative citation orders

The usefulness of a particular citation order depends on the way the collection is used. Clearly, a user who is more interested in everything relating to a particular product than in any other aspect of the subject might prefer to ignore the normal citation order and cite Table 2 first, rather than after Table 0 and 1, so that a document on plaster-board partitions might be classified as Rf7 (22) rather than (22) Rf7. Another user might prefer to cite the material first, giving f7R (22). CI/SfB notation, by giving each major facet a distinctive symbolism, has the great merit of readily allowing different citation orders.

Standard citation order should, however, be used for all published literature and bibliographies, and whenever there is no pressing demand for a tailor-made one. Where such a demand exists, it seems sensible to use a local modification. Although the making of class numbers with a different citation order is perfectly straightforward, two points should be observed by any user following a citation order other than the standard one:

1. Having settled on a given citation order, observe it consistently; predictability is a vital pre-requisite for effective retrieval and is only achieved by strict adherence to a definite plan of campaign.
2. The filing order (strictly speaking the filing order between facets) is best left unchanged. The ordinal value of SfB symbols should be the same in all indexes if users are to be at ease with all SfB files.

2.37 Summary classes

Appendix 2.32 implies that the basic question is whether the document deals with a particular building type, a particular element, etc. If not, the facet in question is regarded as 'not present' and ignored. However, the situation is not always as simple as this. There are several different levels of generality possible in referring to the concepts in a facet. If the subject of acoustic properties of building materials is taken as an example, a document could refer to the materials concept in the following ways:
1. It might not even mention it explicitly. A book called Acoustics in Building might refer to acoustics in different materials and so on. These facets are 'diffuse'. Such a document would not get the class number for materials in its class number at all. Nor would it get the number for any

other particular facet, except that of properties to get the number for acoustics in general: (P).
2. It could refer explicitly to a whole facet, eg: acoustic properties of building products. In this case it would get the class number for building products in general as well as the number for acoustics: Y(P).
3. It could refer specifically to just one product – and would get the class number for this, eg: Ri4(P) acoustic properties of laminated wood sheets.
4. It could refer to two specific products, eg: 'rigid sheets and sections' without putting more emphasis on one than the other, and would get the class number A(P).

An important distinction made throughout CI/SfB, particularly important for project information applications, is that made between summarizing in the sense of one thing plus another, eg: water supply services and gas supply services, which is covered by the nearest preceding summary class – (in this case (5–); and summarizing in the sense of some characteristic common to (pertaining to) both classes, normally placed at the last occurring code in the relevant sequence, in this case (59).

2.4 Filing

2.41 Recommended order

The recommended filing order follows the order of the classes as set out in the tables. It was shown in Appendix 2.25 and another example follows:

Table 0
8	Residences
81	Housing
81(24)	Stairs
81(24)Xf(J)	Precast concrete components – strength
81Yg	Ceramics
81(P)	Acoustics

Table 1
(21)	Walls – external
(24)	Stairs
(24)Xf	Precast concrete components

Tables 2 and 3
Ng	Overlap tiles – ceramic
Ng(G)	Overlap tiles – ceramic – appearance
Xf	Precast concrete components
Yg	Ceramics (as a material)
Yq	Concrete (as a material)
Yq(P)	Concrete (as a material) – acoustics

Table 4
(G)	Appearance
(P)	Acoustics

A possible weakness of this is that it fails to keep the general before the special, eg stairs for housing precedes stairs general.

Appendix 2: Classification and filing

2.42 Inverted order

Collections wishing to observe general before special consistently must follow an 'inverted' order in filing, where the Cl/SfB symbols are used in the increasing ordinal value of (A)/(Z), a/z, A/Z, (0)/(9), 0/9. All these would file **after** 'nothing' (a space). As an example:

Table 4
(G)	Appearance
(P)	Acoustics

Tables 2 and 3
Ng	Overlap tiles – ceramic
Ng(G)	Appearance
Xf	Pre-cast concrete components
Yg	Ceramics (as a material)
Yq	Concrete (as a material)
Yq(P)	Acoustics

Table 1
(21)	Walls – external
(24)	Stairs
(24)Xf	Pre-cast concrete components

Table 0
8	Residences
81	Housing
81(P)	Acoustics
81Yg	Ceramics
81(24)	Stairs
81(24)Xf(J)	Pre-cast concrete components – strength

The practical effect of this order can be seen from the items in class 81 houses. In file 81 or class 81 on the shelves, broad items such as houses precede less broad ones such as stairs in houses and these precede even more specific ones like the strength requirements of pre-cast concrete stairs for houses. The files will not be in the same order as the tables in the Manual, and initially at least, some people will find difficulty in understanding what has happened. In the Manual the tables are given in the order 0, 1, 2, 3, 4, but on the shelves the items classified by Table 4 Activities and requirements will come first and building types in Table 0 Physical environment will come last.

2.43 Other orders

The two methods described are not the only possible ways of arranging documents or items classified by Cl/SfB. The five tables could, for example, be arranged in the order, Table 1, Table 2, Table 3, Table 0, Table 4, or in any order if this was thought essential to the collection concerned and helped users. In most cases this would not be justified. However, filing order between facets is not of great importance and Cl/SfB keeps the order it has because it is probably somewhat simpler than 'inverted' order.

Appendix 2: Classification and filing

2.5 Auxiliary signs and filing order

2.51 Relation sign (colon)

: (colon) may be used to introduce the second class where two classes from the same table are quoted and where brackets are not used, ie:

376:96	Prisons – storage facilities
Fg2:Xt6	Clay brickwork – fixing components
but	
(13) (56)	Floor beds – heating systems
(F4) (M)	Dimensions – heat

This symbol does not have quite the same meaning as the UDC colon, since in most cases in CI/SfB it implies an order of precedence, the 'major' idea being given first. Note that special care in interpretation is needed if codes from the list of common element parts (Appendix 1) are used in conjunction with Table 1, eg:

(57:53)	represents Air conditioning services – grilles (Table 1 and common element)
(57) (53)	represents Air conditioning services – Water supply services (two Table 1 elements)

2.52 Aggregation signs (stroke and plus)

The idea of summary classes is well established in CI/SfB for both project and library applications, but, instead of using simple main class numbers, it may sometimes be helpful to use the stroke and plus signs to give more precise classification.

+ may be used to link non-consecutive class numbers, eg:

31 + 33	Administrative facilities plus commercial facilities (or use 3)
(55) + (57)	Space cooling services plus air-conditioning services (or use (5–))

/ may be used to link the first and last of a series of consecutive class numbers, eg:

31/33	From Administration facilities to commercial facilities, including 32 Offices (or use 3)
(55)/(57)	From space cooling services to air conditioning services, including (56) space heating services (or use (5–))

Note that (55) + (57) and (55)/(57) would normally precede the simple class (55) in a filing sequence, but for some applications it may be desirable to file them after (5–). The policy adopted will depend on the needs of users and should be followed consistently.

In the example of filing order below the colon files immediately after the simple class number. The stroke and plus signs file before it:

31/33	Administrative facilities to commercial facilities, including 32
31 + 33	Administrative facilities plus commercial facilities, excluding 32
31	ADMINISTRATIVE FACILITIES
31:93	Administrative facilities – kitchens (eg: design of kitchens in administrative buildings)
311	International administrative facilities (eg: UN building)

3 Indexing

Like Appendix 2, this section is based on 'The organisation of information in the construction industry' by J. Mills FLA.

3.1 Simple alphabetic subject indexing

Inevitably, if information on any one subject is kept together in a classified collection, information on **all the others** will be scattered to some extent. Alphabetical indexes, besides acting as a key for the location of classes, also help to overcome this scatter by bringing references together. They make it much less likely that information items will be overlooked.

Figure 28
Indexes help users find information which would otherwise almost certainly be missed. This example is from RIBA Product Data. The user is much more likely to find C (Construction) Note 74/20 because it is entered in the index under both 'Concrete breaking plant' **and** 'Demolition'.
An item on bathroom fans may be filed on the shelf under 'Bathrooms' at 94 or 'Fans' at (57), so may easily be missed by users who are not expert librarians. But the index can list it under **both** terms.

```
Concrete, see                          Decorative coatings: internal finishes, thin (42) V
  Aerated concrete                       :wall/floor finishes, thick (4—) P
  Dense concrete                       Deep-fryers: cooking equipment—C Note 75/15
  In-situ concrete                     Defects: liability
  Lightweight aggregate concrete         :in England—P Note 76/1·023
  Precast concrete                       :in Scotland—P Note 76/1·024
  Prestressed concrete                 Delegation and control—P Note 76/1·044
  see also                             Demolition
  High alumina cement                    :concrete breaking plant—C Note 74/20
  Pfa                                    :removal of asbestos lagging—C Note 75/112
Concrete accelerators: alternatives to calcium   Demonstrations: architects—P Note 75/76
  chloride—C Note 75/113               Demountable partition systems (22·3) X
Concrete admixtures Yu2                Dense concrete: blocks Ff2
Concrete admixtures/adhesives Yt3      Deontology, see EEC Agreement on Deontology
Concrete breaking plant—C Note 74/20   Department of the Environment
Concrete bricks Ff2                      :fluctuations/fixed fee contracts—P Note
                                          75/101
```

Each document or set of documents should ideally have its own alphabetical subject index. If more than one facet is used, eg Table 0 as well as Table 1 in CI/SfB, with detailed subdivision rather than just main headings, then the compilation of a subject index becomes a practical necessity if retrieval is to match filing efficiency. The cost of indexing should be taken into account when deciding how much classification detail to use.

Indexes in office libraries have to be modified each time the library is updated or the classification changed, and this can increase their cost. But a good index saves considerable time and thus money to its users. Without it, a lot of information will be missed. In short, alphabetical indexes should be provided to classified collections of information whenever the chosen level of detail requires this.

3.11 Constructing an indexing

To construct an index, decide what subjects and hence what terms to index for each item (document or paragraph), bearing in mind the user's likely approach. Make an entry for each subject and write the entry and corresponding code on a library card (usually 5''   3'' or A6 size plain card). At a later stage, sort the cards, so that the entries are in one alphabetical sequence. Because thousands of index entries could be created for a very small number of documents, it is essential to find some simple but systematic method of keeping the number of entries in an index down to a minimum without reducing its efficiency. Chain indexing does this, and has two advantages:

1. It ensures that every significant keyword for each subject appears in the front position, so that it is in the alphabetical sequence.
2. It does this very economically.

Appendix 3: Indexing

An example:
Assume two documents are to be indexed, one on **floors for housing**, the other on **acoustics of floor finishes for housing**, and that the classification is available for each:

1. Write the analysis for the first document down on a work sheet:
Housing 81 – floors (23)

This is really a shorthand method of representing the hierarchy or 'chain' in which CI/SfB places the subject:

[Residences   8]
  Housing   81
    Structure   (2–)
      Floors   (23)

2. Write down index entries for these four subject terms, on four separate cards, starting with the one furthest to the right, ie (23) Floors and moving to the symbol furthest to the left, ie: 8 Residences. Each item is qualified by such of the others, working right to left, as seems necessary. The entries are:

Floors: housing   81(23)   (omitting 'structure' and 'residences')
Structure: housing   81(2–)   (omitting 'residences')
Housing   81   (omitting 'residences)
Residences   8

The qualification of 'floors' and 'structure' by 'housing' is essential because the information which is being indexed is **only** to do with housing. Information on floors for other buildings, eg: hospitals is not at the numbers given.

3. Write the analysis for the second document down on a work sheet:
Housing 81 – floor finishes (43) – acoustics (P)
representing

[Residences   8]
  Housing   81
    [Finishes to structure (4–)]
      Floor finishes   (43)
        Acoustics   (P)

4. Write down index entries for these five subject terms starting with the one at the right hand end, ie: (P) Acoustics, again qualifying each by such other terms, working right to left, which are necessary. The entries are:

Acoustics: floor finishes: housing   81(43)(P)   (omitting 'finishes to structure' and 'residences')

Floor finishes: housing   81(43)   (omitting 'finishes to structure' and 'residences')

Finishes (to structure): housing   81(4–)   (omitting 'residences')
Housing   81
Residences   8

The entries for housing and residences are not needed in practice because entries have already been made (in stage 2 above) for these subjects. The price paid for this economical method of indexing is that users have to remember that they will find the most direct and helpful references under the most specific subjects, eg: Acoustics (of) floor finishes (for) housing 81(43)(P).

Appendix 3: Indexing

5. Rearrange all the entries in one alphabetical sequence, ie:
Acoustics: floor finishes: housing	81(43)(P)
Finishes (to structure): housing	81(4–)
Floor finishes: housing	81(43)
Floors: housing	81(23)
Housing	81
Residences	8
Structure: housing	81(2–)

It may be objected that housing has been indexed:
Housing   81
when there is nothing on housing in general. There are two answers to this: first, there certainly is something on housing; and, as the collection grows, more material is almost certain to be acquired on these general classes; secondly, the main function of any given A/Z index entry is not to index a particular document, but to lead the user to that part of the library where the documents on the subject in question begin. The real meaning of the simple entry **Housing 81** could be elaborated in the following terms: 'housing: information on this subject begins at 81 and particular sub-classes and aspects of it will be found following it'.

A number of finer points suggest themselves when making an alphabetical index. An important one is providing for synonyms. A good indexer, as he considers a term for possible indexing, always asks himself if it has possible synonyms which people may look under. If there are any, they should be given 'see' references of the kind 'sound *see* acoustics'.

3.2 Post co-ordinate indexing

Post co-ordinate indexing – usually called simply 'co-ordinate indexing' – does not co-ordinate at all in the sense in which that word is used throughout this manual. The section dealing with project information applications suggests that information contributed by different members of the building team can be more effectively 'co-ordinated' (related), to the advantage of all, if some common rules are accepted for structuring and presenting it. Co-ordinate indexing, on the other hand, allows the producer of information greater freedom to index it so thoroughly and in such detail that it will probably be retrieved whether or not it has consistent structure. Co-ordinate indexing is likely to be unsuitable for most project information applications. On the one hand, structured information has outstanding advantages, eg: for checking completeness; on the other, very detailed indexing is impracticable and uneconomic. The technique is more likely to be useful for the retrieval of general information including 'project-related' information, and for keeping records when it is important to be able to trace items subject to different combinations of circumstances.

An example of this might be personnel records in a large organisation in which it might be necessary to trace all members of staff who eg: live North of London/ are over 25/ have been with the office more than x years; or to select building products according to the way they combine a number of defined and indexed properties. The simplest way to explain it is to demonstrate how a particular application works – in this case the 'feature

Appendix 3: Indexing

card' system, which is the most popular form of co-ordinate index in the UK. There are several other names for this system, including 'peek-a-boo', 'optical coincidence', and 'Batten cards'.

3.21 Feature cards

Every document is given a unique accession number, 1, 2, 3 etc, as it is added to the collection, and the documents are normally kept in accession number order, not according to subject. The index consists of cards which may be 8" × 8" but there are a number of sizes available. Each card represents, not a document, but a simple feature (term or concept), and on the card are recorded the accession numbers of documents in the collection which refer in some way to the concept. If the feature card **door**, for example, has the number 1674 recorded on it, this means that document 1674 deals to some extent with the subject of doors.

Figure 29
Feature cards for coordinate indexing.

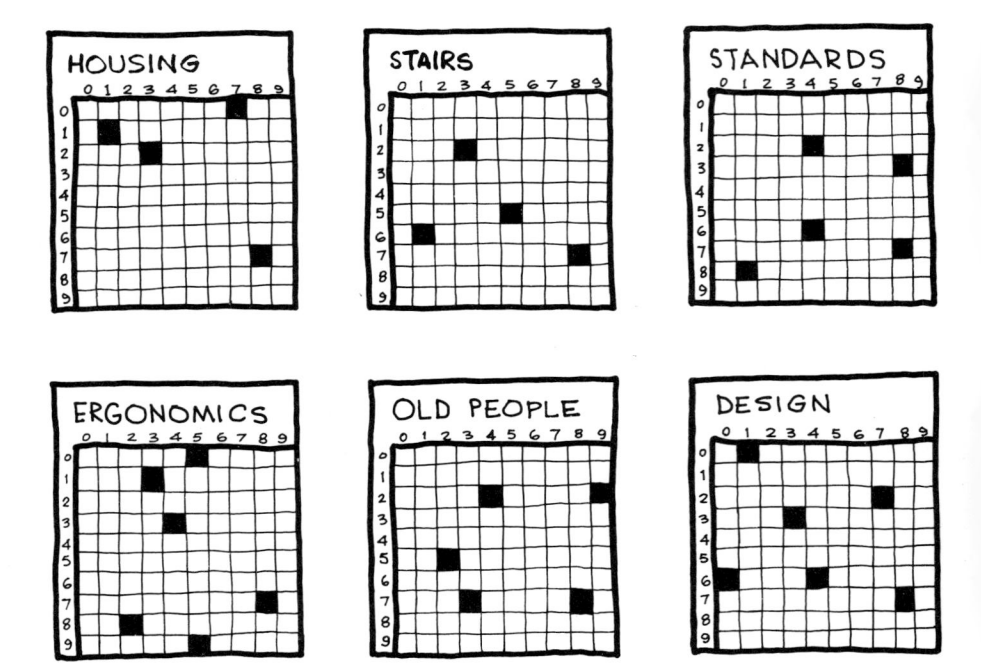

The document number is recorded only as a hole punched in a certain position, as in Figure 29. Each card shows these positions by carrying a grid of small dots, numbered 1–10,000 or whatever is the capacity of the card. For simplicity, only 100 positions are shown on the examples. Documents numbers 7, 11, 23 etc include the term 'house' in their description.

The cards may be kept in A/Z order of terms, or they may be arranged in some classified order, eg: CI/SfB. Note that the terms are kept as simple as possible, although phrases like 'old people', are occasionally accepted.

3.22 Indexing

As in conventional indexing, this has two basic steps:
1. The document is concept-analysed and the indexer decides what concepts make up the subject of the document. This is usually done by extracting from the title, summary and text, the keywords which best convey its subject. How many terms are chosen to represent the subject of the document is purely a matter of policy on the part of the library concerned. It might assign 5 or 6 on an average; many UK special libraries use between 8 and 12.

Appendix 3: Indexing

2. These concept terms are translated into the controlled index language. The document may, for example, refer to domestic staircases but when the two concepts implied ('domestic' and 'staircases') are checked in the index language (often called a thesaurus) they might have to be translated into houses and stairs.

3.23 Locating documents

If a request is received for documents on designing staircases in old people's homes, the keywords which name the subject in the question are translated into the indexing language, eg: design, stairs, old people, houses. The cards for these four terms are extracted from the file, superimposed together carefully, but in no particular order, with all four edges coinciding exactly, and held up to the light.

If any document in the collection has been indexed by all the terms, then a hole would have been punched at the same accession number position for all four cards — and the light would shine through, to symbolize the matching of the search prescription and that document's index description.

From the cards in the example above it can be seen that document 78 was indexed by all four question terms, since that number was punched for all four cards. (It might have been indexed by other terms as well, but that is irrelevant at this point.) It is assumed that document 78 is something to do with design of stairs in old people's homes and it is therefore retrieved and examined. This is easily done if all the documents are shelved in accession number order. If, however, they are arranged in some other way (eg: by Cl/SfB) another simple index must be kept, arranged in accession number order and indicating against each number where the document itself is shelved.

If no light shows through when the four cards are superimposed, this would seem to mean that no document in the collection dealt exactly with the subject of the question. Or, position 78 might show a light, but document 78 might not, on examination, give as much information as the questioner wanted. In these cases, the search would have to be adjusted — the class of documents examined made larger, or broader, to see if any other documents were worth looking at. This could be done in two ways:
1. One or more of the term cards could be left out when searching, eg: the card for design might be dropped. This would make the question less specific and a document would be accepted if it dealt with stairs in old people's homes even though it did not deal specifically with design. Or, the card for old people might be dropped, and so on.
2. Each concept could be broadened individually by examining the term cards for containing classes. The card for residences and the cards for any other likely type of residence other than houses (eg: hotels), could be examined in combination with all or some of the other terms. Or, other members of the class 'special categories of user' (eg: handicapped persons), might be examined. When using this method to adjust the classes to be examined, the arrangement of the cards (the index file) in classified order would be a great advantage.

Appendix 3: Indexing

The term 'co-ordination' is something of a misnomer for the basic operation, which is more correctly called 'intersection' and can be represented thus:
A = the class of documents referring to subject 'A'.
B = the class of documents referring to subject 'B'.
AB (the intersection of A and B) = the class of documents referring to the subjects 'A' and 'B' together.

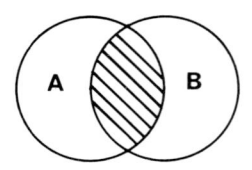

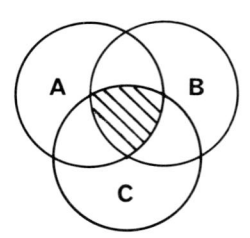

Clearly, the more terms 'co-ordinated' together at the same time (ie: the greater the number of cards placed together), the smaller the resulting intersection and the smaller the class of documents corresponding to the description.

3.24 Advantages claimed for coordinate indexing

Freer access
Whatever combination of terms is co-ordinated to see what documents are available, these documents are immediately identified (at least by their accession number). This may be compared with the conventional catalogue (one arranged by Cl/SfB, say) where many combinations will be found distributed, so that the class represented by the combination of 'housing' and 'old people' might be found under:
Housing – old people
Housing – low rise – old people
Housing – stairs – old people
Housing – floor finishes – old people

The reason for this is that in the conventional catalogue the co-ordination or intersecting (compounding) of terms is recorded in the index at the time of indexing and before any particular question is received. This is necessary in order to produce a shelf order or order of entries in a catalogue. In the co-ordinate index, no co-ordinating of terms is performed until after a question is received. There is therefore no need to subordinate some terms to others, and no scatter or 'distributed relatives'. This distinction has led to the use of the names 'pre-co-ordinate index' and 'post-co-ordinate index' to refer to the two kinds; the latter is the correct name for 'co-ordinate index'.

More exhaustive (thorough) indexing
A conventional catalogue represents the information content of a document by summarization, so that a Cl/SfB number represents a class which sums up the major theme of the document. It could, of course, go further than this and have additional entries for subsidiary themes; a report on swimming pool design, for instance, might have a section on the special problems of concrete construction involved, and this might be given an additional and separate entry of its own. But if this multiplying of entries were done on a large scale, it would seriously enlarge the bulk and maintenance costs of the catalogue.

Appendix 3: Indexing

In post co-ordinate indexing, the translation step is easier since no problems of citation order or of notation arise. Consequently, extra terms are usually assigned more freely. A recent report, for example, was entitled 'Optimum dimensions for domestic stairways'. Translating the title into keywords dimensions, houses and stairways gives an approximate summarization of the central theme of the paper. In co-ordinate indexing, it would be possible to go further and assign as additional index terms the following keywords, all of which appear prominently in the summary and full text of the paper; ergonomics, physiology, psychology, rise, going, standards, experiment.

The main result of such high exhaustivity in indexing is to improve the chances of all the relevant material being found in response to a question: a request for material on physiological requirements in the use of stairways would, for example, at once retrieve this document. It should not be assumed, however, that because a conventional catalogue would not index this document explicitly by the term 'physiology', it would not be found in response to such a question. For it is frequently the case that subsidiary themes are implicit in the summarization. For example, the simplest CI/SfB class for the above document would be houses – stairs – dimensions. This assumes a policy of only one class number per table being accepted. If more precise descriptions were allowed (by citing more than one concept from the same table where necessary), the CI/SfB class would be houses – stairs – dimensions – ergonomics (both dimensions and ergonomics coming from Table 4).

If careful broadening of the search takes place, a search for physiology in relation to stairways would include looking at the containing class, and this is ergonomics. Similarly, a request for information on 'rise' or 'going' specifically would recognise that these are properties of stairways and hence subordinate to stairs; or, if a request was made for information on experiments relating to stairs, it would be recognized that any method of studying a problem will be subordinated to that problem. In each case, the document would be found; but the locating of it might not be quite as prompt as with an exhaustive co-ordinate index.

Ease of mechanization
It was shown that the presence or absence of a hole indicated whether a particular document had or had not been indexed by a particular term; also, that a beam of light was used to recognize the fact – a light visible represented a 'yes', and no light represented 'no'. This demonstrates that once the indexer and searcher have set the scene (a file of index descriptions and the receipt of search prescriptions to be matched against them), the matching itself is a purely mechanical operation and a variety of mechanized methods can be used to do this matching, from a photo-electric cell to a computer. Consequently, co-ordinate indexing is often linked with mechanization and this is often assumed to be a particular advantage.

It is important, however, to see that all that is mechanized is the final matching operation – not the concept analysis of a document, or of a question, nor the translation step into a controlled language. And if a machine matching operation is compared with a human indexer searching an ordered index file, the latter is found to be a remarkably efficient machine. This is not to say that mechanization of a co-ordinate index does not offer advantages – only that it is important to know just what it is doing.

3.25 Use of CI/SfB in coordinate indexing

It has been assumed that keywords extracted from a document have to be translated into some controlled indexing language. Although experiments are being made to test the possibility of omitting this precautionary step in co-ordinate indexing and relying almost entirely on 'natural language', it is wise to assume that a fair degree of control is desirable. In a number of British and American co-ordinate indexing systems, this controlled indexing language takes the form of a 'thesaurus' of terms, alphabetically arranged. The thesaurus for the UK construction industry is the 'Construction industry thesaurus' (CIT).

So far as possible, the terms used in this edition of CI/SfB have been related to the terms used in CIT, although they sometimes assume a different level of importance. There may well be cases where it may be convenient to use CI/SfB terms as keywords for post co-ordinate indexing.

4 Theoretical basis of CI/SfB

There are inevitably fringe areas of application for all classification systems and many people will have good reasons for feeling uncertain as to the usefulness of CI/SfB for their purposes, or will only be able to make partial use of the scheme. This last section of the Manual is addressed to them. It explains the nature and objectives of the system and the more obvious limitations on its use which follow from these. Lastly, it discusses the circumstances in which partial use of the system may be practicable.

4.1 Nature of the system

CI/SfB is primarily a classification of buildings and their parts. The basis for the arrangement of the schedules, from the beginning of Table 0 to the end of Table 4, is that parts follow the things and types of things to which they can reasonably be said to relate. In the example of the sequence given below, block work is listed in the system after external walls because it can be regarded constructionally as part of external walls, and external walls are listed after schools because they can be regarded as parts of schools. (The reverse is not possible: schools cannot be regarded as parts of external walls!)

Thing eg	School	Table 0	Code 71
Types eg	Primary school	Table 0	Code 712
Parts eg	External wall (may be part of primary school)	Table 1	Code (21)
eg	Blockwork (may be part of external wall)	Table 2	Code F
eg	Blocks (may be part of blockwork)	Table 2	Code F
eg	Aggregate (may be part of blocks)	Table 3	Code p
Other factors eg	Cost	Table 4	Code (Y)

Appendix 4: Theoretical basis of CI/SfB

The example also shows how non-objects follow (in Table 4) the objects listed in Tables 0, 1, 2 and 3. In the words of Augustus de Morgan, the sequence is one in which 'Great fleas have little fleas upon their backs to bite 'em. And little fleas have lesser fleas and so ad infinitum'. The corollary of this is that CI/SfB can be looked at, in reverse order, as a system of 'levels of aggregation', a model for the building process in which each level adds to what has gone before until the final result is achieved:

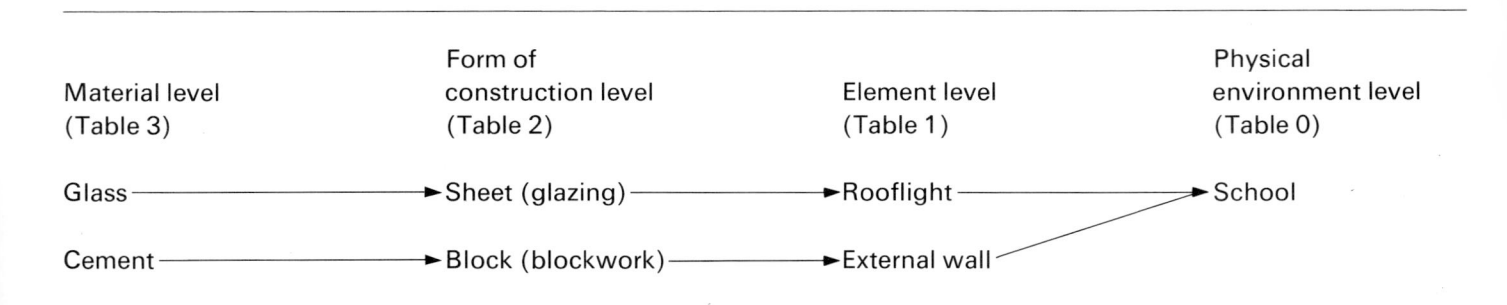

Material level (Table 3)	Form of construction level (Table 2)	Element level (Table 1)	Physical environment level (Table 0)
Glass	→ Sheet (glazing)	→ Rooflight	→ School
Cement	→ Block (blockwork)	→ External wall	→ School

4.2 Objectives, limitations on use

The word 'Construction' in the title of CI/SfB is appropriate in two senses. First, because the system is limited to the construction industry. It makes no claim to be suitable for use by practitioners in other industries. Second, because it adopts the construction view-point in the construction industry. It derives primarily from the 'levels of aggregation' as they occur in the total construction process, and seeks to arrange each category in a way which seems sensible to practitioners, to produce a classification of practical value.

As examples of this, Table 0 gives prominence to (that is puts first) characteristics which most clearly identify classes in the eyes of practitioners, eg: 'primary school' rather than 'state school'. Also, Table 1 orders its major classes in a way which relates broadly to the construction process. Substructure comes at (1–) before (2–) Structure, the shell of the building built on the substructure. (8–) Loose fittings, representing items which are moved into the buildings at the final stage, comes at the end of the table.

Construction is the process of assembling or aggregating resources, particularly materials. But before this can take place on site, or (in the case of manufactured articles) in a work-shop, a great deal of information is required on the physical resources needed, on the actual work of assembly, and on the end results to be achieved. **Resources** have to be selected, specified, drawn, measured, bought. **Assemblies** (of resources assembled in their final position on site) have to be specified, drawn and measured. **End results** have to be planned in terms of design and cost.

Of these activities, selection requires product information libraries, both national collections and those made specially for individual offices and projects; drawing requires libraries of type drawings and drawings prepared

Appendix 4: Theoretical basis of CI/SfB

specially for individual projects. Similarly for specifications and measured items. CI/SfB provides a common, consistent means of subdividing primary groupings of end result, assembly and resource information, as follows:

End result information*	Assembly information* (actual work of construction)	Resource information* (resources needed for construction)
CI/SfB concepts and codes in their standard order, eg: **(23) Floors** **(24) Stairs** provide a consistent means of subdividing information on finished work in, eg: Cost planning information; Regulations; Performance standards; Location drawings; etc	CI/SfB concepts and codes in their standard order, eg: **(23) Floor construction** **(24) Stair construction** provide a consistent means of subdividing information on assemblies, eg: Construction planning information; Workmanship specifications, measured items; Codes of practice; Assembly drawings; etc	CI/SfB concepts and codes in their standard order, eg: **(23) Floor components** **(24) Stair components** provide a consistent means of subdividing information on resources in, eg: Commodity information; Material specifications; Schedules; British Standards; Component drawings; etc

*These three major categories are not usually given codes because they commonly form clearly identifiable groupings of information, well separated from one another. However, sets of production drawings (see Product information 1.3) can often include all three and individual drawings may be coded L (location, 'end result' drawings); A (assembly drawings); or C (component, resource drawings).

The method of using CI/SfB for subdivision of particular primary groupings of information is bound to vary slightly but this will usually be preferable to arranging each separate document or collection by ad-hoc or special methods entirely different from those used for other primary groupings.

The principles underlying the development of CI/SfB are stated briefly below. The statement is presented in terms which will be more familiar to classificationists than to building practitioners and is based on notes prepared for discussion between the SfB Agency, the Data Co-ordination Secretariat of the Department of the Environment and Jack Mills, consultant to the RIBA and co-director of the CIT project.

CI/SfB provides a common arrangement (sequence of topics) for project documentation (PD) and related general documentation (GD), but gives the needs of PD priority over GD. Within PD, it gives the needs of the assembly function (the actual business of constructing) priority over other functions.

Appendix 4: Theoretical basis of CI/SfB

The need to achieve improvement of building design from project to project suggests that the brief/feedback link needs to be strengthened in building practice. The most realistic way to do this is to use a single classification for GD and PD. The existence of a large body of 'quasi-PD' literature (eg: national specifications) reinforces the need for a single classification for GD and PD.

Functions of retrieval languages
BC Vickery has summarised the functions which retrieval language might perform, and emphasises that no one language would be expected to perform all of them. They are:
1. To provide for the selection of bibliographic records with unique characteristics – for example, any record containing a specified string of juxtaposed words, such as 'enzyme activity in mammalian cells'. (Selecting by specified word-strings, implying a computerised, natural language system.)
2. To provide for the selection of a set of records likely to be relevant to a particular topic – the function of specific reference (specific search).
3. To provide for the selection of a group of records lying within a certain subject field – for example, in an agricultural and biological index, all items concerned with Parasitism. This is the function of generic survey, (generic search).
4. To provide for the sequencing of a set of selected records according to probable relevance to a particular topic – by means of some ranking device. (A system of ranking retrieval items by probable relevance, usually associated, like 1, with mechanical systems only.)
5. To provide for the ordering of a group of records in a field into a meaningful sequence – for example, to arrange by subject the selected items concerned with Parasitism, (systematic arrangement).
6. To provide for the automatic conversion of index entries from one form to another (eg: by chain procedure or by rotation).
7. To provide for conversion from one language to another (eg: through use of the A/Z index or by providing different sortations of items to meet the needs of different users).
8. To give aid to the searcher in his choice of search terms (aiding search programming).

CI/SfB, like all consistently structured 'library classification' type languages, performs a large number of these functions. Functions 2, 3, 5 and 8 are all traditional functions of a classified index, and can be carried out by CI/SfB. Functions 6 and 7 are less obviously met by CI/SfB but can be met to a greater or lesser extent. Functions 1 and 4 are unlikely to be within the capability of a fully-operational CI/SfB index. In relation to building process functions, PD usually requires much less specificity in vocabulary than GD and involves the organisation of a smaller universe of concepts and terms. Because of this, the functions of retrieval languages listed above, though equally applicable in both situations, are more critical in GD than PD. Those which seem most relevant, in the experience of the SfB Agency, to the common needs of building industry practitioners, are **coordination and cross reference** (implied by 2 and 3), **systematic arrangement or search pattern** (5), **consistent titling and item headings** (8), and **sortation** (7).

Appendix 4: Theoretical basis of CI/SfB

Additionally, CI/SfB as an operational system aims: **to be maintained and improved** from edition to edition so as to provide its users with the best widely available and commonly accepted tool for their purposes; **to be reasonably stable**, ie. not be revised unnecessarily frequently or drastically. A particular problem here is to reconcile the needs of new users – who want the best possible and most up-to-date system – with existing users who inevitably suffer the burden of changing from one edition to the next; **to be presented in a manual** or series of manuals structured so that they can readily be understood and used by practitioners, complete with alphabetical indexes and procedural notes on the application of the system to each relevant function; **to enjoy the widest possible measure of national and international support**, putting national views before international views only in the last resort where these cannot be reconciled.

The effectiveness of coordination with CI/SfB is dependant on the extent to which the system is used for PD and GD building process functions and on whether each individual function uses it in a reasonably similar way. A clear statement of how the system is to be used for each function is of great importance in project and office situations.

4.3 Partial use

Four cases of partial use

1. Organisations within the construction industry but have to cater for specialised information in considerable depth, will almost certainly find that CI/SfB is inadequate for their specialism. They may be able to use the system for all areas except their specialist area and will need to provide their own classification to operate in a CI/SfB framework.

2. Organisations may have a construction industry viewpoint, but may also have substantial ancillary information on other subject areas, eg: medicine, philosophy, etc. Such organisations will need to make extensive use of the peripheral classification (based on UDC) or adopt an alternative. If they use a general scheme as a peripheral classification, they may need to use its schedules in far more detail than is provided by the UDC schedules at (Z) in this manual.

3. An organisation may be within the construction industry but have a point of view quite different from that adopted by CI/SfB. It may, for example, require an arrangement of information which emphasises architecture as an art form. In such a case, it may be possible to allocate CI/SfB a subordinate role within a specially constructed frame-work more suited to the needs of the members of the organisation. That is to say, it may be possible to subdivide some schedules, particularly those most relevant to practical construction, by means of some parts of CI/SfB.

4. Perhaps the most difficult case of all, is that of the organisation which is half in and half out of the construction industry. So much will depend on the individual case that it is almost impossible to give general advice but if a reasonably high proportion of the members of such an organisation are involved or will eventually be involved in the practical work of designing, constructing or maintaining buildings, CI/SfB should probably be used, in whole or in part.

5 History of SfB

Sections 5.1 to 5.3 are based on an article by Brenda White ALA which first appeared in the Library Association Record in December 1966; Section 5.4 brings the account up to date.

5.1 Classification for the building industry

In 1947, the authorities responsible for rebuilding destroyed areas in Belgium and France after the Second World War called for a conference on building documentation, to be held in connection with the Paris International Exhibition on Housing and Building in the summer of that year. The most important topic at the Conference was the international need for documented information, and the organization of such information.

The 1947 Conference, taking as a model of a documentation service, Finland, where SAFA Building Standards Institute (Note 1.1) published information sheets preclassified for filing, agreed that the format and classification of documents for filing should be standardized.

Their recommendations were first, the format to be the international A4 size, and second, for classification purposes, a rectangular box (45 × 20mm) to be printed at the top righthand corner of the first page of each document, the box to be equally divided by a horizontal line, the lower space to be occupied by the appropriate UDC number, the upper by a simpler notation (not specified by the Conference) by which the document should be filed. The problem of an international classification, however, was left unsolved.

In 1952, the International Federation for Documentation (FID), and the International Council for Building Documentation (CIBD, later CIB) set up a joint committee to investigate this problem: the International Building Classification Committee (IBCC). The work of the IBCC was set in four phases:
1. To study and publish selected UDC numbers to be used for building classification.
2. To study and publish the Swedish SfB filing system.
3. To study other systems of classification and filing in the building field.
4. To develop a standard method for classification and filing.

5.2 Background

The work of the IBCC began in 1953, and Phase 1 resulted in the publication in 1955 of ABC:Abridged Building Classification for Architects, Builders, Civil engineers, a selection from the Universal Decimal Classification. This had been developed from a similar selection of building headings from UDC made by Agard Evans for the Ministry of Works Library, and published by the Ministry in 1945 (corrected and expanded 1946).

Work on Phase 2 began in 1955. The SfB system had been studied by experts in 1949 and 1950. A report of the system by Egil Nicklin, director of the Finnish Building Standard Institute, was studied by the IBCC, and subsequently published in 1957 as IBCC Report No.1. Phase 3, a comparative study of filing systems already in operation throughout the world began in 1957. Fifty-five systems were received and compared; the conclusion reached was that the two most useful systems in operation were

---

5.1   Now known as SAFA Stadsplane och Standardiserings-institutet (SAFA town planning and standardisation institute).

Appendix 5: History of SfB

UDC, for its wide subject coverage, and SfB, for its shortness, flexibility and relevance to building practice. This survey was published as IBCC Report No.2 in 1959. An SfB Committee, consisting of members who used the system, was set up at the first CIB Congress in Rotterdam in 1959 and made responsible for future development.

The same conclusion was reached by the British architect, Dargan Bullivant, working in his capacity as Research Fellow on 'Information for the Architect' appointed by the London-published weekly magazine, the *Architects' Journal*. In studying the requirements of a standard classification, Bullivant made detailed reports to the *Architects' Journal* research board on UDC, SfB and the American Institute of Architects Standard Filing System. Dargan Bullivant was also a member of the IBCC working team which investigated Phase 3 of the Committee' work. The team's report on basic problems of classification and filing is published as IBCC Report No.4, and Bullivant himself wrote the team report on filing according to a complementary SfB/UDC system, IBCC Report No.5. This system was also published in September 1959, by the *Architects' Journal* as a Building Filing Manual.

The team reports were presented collectively to the sixth plenary meeting of the IBCC in 1959. Accepting the conclusion that the most suitable system for the building industry was SfB, the IBCC made the following recommendations:
1. That CIB should publish or promote the publication of the SfB system in several languages for use in the classification of trade catalogues, codes of practice etc.
2. That copyright be vested in CIB.
3. That the SfB tables should be amended only on the advice of the IBCC.
4. That SfB/UDC building filing manuals might be published nationally on the responsibility of CIB member institutes.

In Great Britain, copyright of, and responsibility for administration, control, and promotion of, the SfB system rests with the Royal Institute of British Architects, which is an associate member of CIB; and in accordance with the fourth recommendation above, the Institute published in 1961, the *SfB/UDC Building Filing Manual*, subtitled 'recommendations for standard practice in preclassification and filing'.

The English version included sections which had no place in the original Swedish SfB system, but which had been added to make the system comprehensive enough for use as a library classification for building practitioners. It also included UDC numbers, both as an alternative classification and as a means of sub-division where SfB itself provides none. The original SfB system, the International Basic Tables, was controlled by the SfB Committee of the IBCC. Both the IBCC and the RIBA guaranteed that no changes would be made to the tables of either version until at least 1965.

5.3 The SfB system: origin and use

The system is Swedish in origin. In 1947, 32 central bodies in Sweden formed a committee for purposes of co-ordination. The committee, which represented all major interests in the building industry, was called Samarbetskommittén för Byggnadsfr Çgor, from which the letters SfB are taken. After three years' work, the committee published in 1950, the Swedish Bygg-AMA (General Material and Work Specification for building), a collection of codes of practice arranged by the system now known as SfB. The principal author of the Bygg-AMA and of SfB was architect Lars Magnus Giertz, secretary of the SfB Committee and Technical Director of the Central Office of the Swedish Association of Architects. SfB was also used for the arrangement of Svensk Byggkatalog, a collection of data sheets on building materials and components issued by the Swedish Building Centre; and Aktuella Byggpriser, a set of loose-leaf sheets on rates and prices of all kinds of building work, issued by the Central Office of the Swedish Association of Architects. A small booklet explained the system of arrangement and showed how adoption of such a standardized system could simplify the whole process of building.

The use of the SfB System in these three important publications led to its gradual acceptance in Sweden and its use as the standard method of filing documents in architects and other practitioners' offices. From Sweden, it spread to the other Scandinavian countries, and elsewhere.

The introduction of SfB in the *Architects' Journal* of 17 September 1959 ensured a wide currency for the system among British architects. After the publication in 1961 of the RIBA manual (the English version), this was also printed in the *Architects' Journal* in September 1961, thus providing most British architects with their own copy of the system, which many used to arrange their personal collections of information.

During 1964 and 1965 the *Architects' Journal*, which used SfB to classify all its technical articles, published an extensive series of articles on Co-ordinated Building Communications (CBC), a computer-operated project management and costing system based on SfB but invented and owned by Bjørn Bindslev, a Danish architect. The series roused considerable interest amongst British architects and quantity surveyors. At about the same time, in April 1965, a symposium on the use of the RIBA manual in library practice was held at the RIBA, under the auspices of the Building Industry Libraries Group (Note 5.2). The speakers' papers and the ensuing discussion clearly indicated the need for revision.

Following this meeting and the description of CBC which suggested the potential for project management and the computer production of interrelated bills of quantity, specifications and networks, the RIBA decided to start work on the revision of the RIBA manual.

By this time SfB was being used for filing purposes by the main product information systems and in 1966 the RIBA, RICS and NFBTE made an investigation to find out what proportion of their members were using SfB.

5.2 Now known as the Construction Industry Information Group.

The results, which suggested that up to 65% of architects, 55% of quantity surveyors and 45% of contractors were using it, were published in the *RIBA Journal* and elsewhere.

Also in 1966 the RIBA issued a report prepared by John Carter, embodying proposals for the revision of the SfB tables. The Technical Committee of the RIBA accepted the report in May 1966, as the RIBA contribution towards the development of SfB, and made proposals for revision based on it to the IBCC in June 1966. Some of these proposals were accepted by the IBCC at two meetings in 1966, and the international tables, as they stood in December 1966, formed the basis for the second edition of the RIBA Manual, published in September 1968 as the *Construction Indexing Manual*, under the RIBA SfB Panel with Dargan Bullivant as Chairman. This, like the first edition, was written principally as an aid to the arrangement of office libraries.

However, a booklet published at the same time as the manual included examples of the use of SfB for the organisation of project documents, as well as an extended account of its use for general information by Jack Mills FLA, the RIBA's consultant classificationist. The version of SfB set out in these publications was named CI/SfB (Construction Index/SfB) to distinguish it from the SfB international Basic Tables in CIB Report No.6.

Following publication of the new manual, the technical press and most product information systems adopted CI/SfB and the RIBA set up a Reference Service to provide publishers of trade and technical literature with CI/SfB classification numbers for their documents. By 1969 a number of architectural and quantity surveying offices had experience of using CI/SfB or CBC for project information and in April 1969 the RIBA Council recommended architects to use CI/SfB for this purpose.

The RIBA asked that advice on project applications should be published and an SfB Working Group was set up with W. R. V. Ward as Chairman to pool available experience. The CI/SfB Project Manual, produced under the guidance of the Working Group, was published early in 1971.

5.4 Data Coordination

From 1968 onwards the Ministry of Public Buildings and Works (later the Department of the Environment) undertook large scale studies of what now became known as data coordination, seen as a means of increasing efficiency and reducing costs in the industry. The Committee on the Application of Computers in the Construction Industry (CACCI) commissioned a study from the Building Research Station which led to the National Consultative Committee (NCC) setting up in 1969, a Working Party on Data Co-ordination which itself reported in 1971, more or less concurrently with the publication of the CI/SfB Project Manual. The Working Party made a substantial number of recommendations for further work and reported on the need for convergence of existing systems. One practical result of the Working Party's efforts, important for the standardisation of terminology and classification development, was the production of the Construction Industry Thesaurus (CIT).

Appendix 5: History of SfB

During this period, CIB acted as a forum for increasing international interest in data coordination and held international symposia in Oslo in 1968 and Rotterdam in 1970. Classification was seen as one aspect of this subject and responsibility for SfB development at an international level, previously undertaken by the SfB subcommittee of IBCC, was given over to the new SfB Development Group, CIB Commission W58.

In 1969, the RIBA established RIBA Services Ltd, following completion of a market survey which showed that architectural offices wanted a service organisation to provide them with aids to practice. The new company gave house room to the SfB Agency UK Reference Service. The CI/SfB Project Manual was serialised by the *Architects' Journal* in 1970 and published in book form in 1971. In May 1972 RIBAS launched RIBA Product Data, the CI/SfB-arranged UK equivalent of Svensk Byggkatalog.

Also in 1969, following a study carried out by John Carter, the RIBA accepted the invitation of the National Economic Development Council (NEDC) to produce a National Building Specification (NBS) for the construction industry. A special company was established for this purpose and the DOE and Greater London Council (GLC) seconded staff to it. After careful consideration, its all-industry board decided that the NBS, the UK equivalent of the Bygg-AMA, should be arranged by CI/SfB, and it was published in March 1973.

A study by the Building Research Establishment (BRE), also published in 1973 (as Current Paper 18/73), showed that CI/SfB could be used with advantage for the arrangement of sets of working drawings. The BRE work reached a wider audience with the publication of *BRE Digest 172* 'Working drawings' in December 1974. Developments in the UK 1969 to 1973 centred on the NCC Working Party's report, the CIT, the NBS and the BRE work, but practical work was being undertaken in many other countries. For example, in Ireland, a government requirement was introduced that publicly-financed building projects should be budgeted for on the basis of cost plans prepared in accordance with SfB Table 1. CI/SfB was translated into French (SI/SfB) and German (BRD/SfB). CIB Commission W58, meeting in 1970, started to revise the SfB Basic Tables, having as a principal aim the eradication of most of the coding and terminological differences between the SfB Basic Tables, CI/SfB and CBC. Following completion of this work CIB Report 22 (the new SfB Basic Tables) and this manual now form the most up-to-date statements on SfB available in the English language. The title of Report 22 'The SfB system: authorized building classification system for use in project information and related general information' marks CIB's endorsement of the use of SfB for project information purposes.

6 Correlation with other systems

6.1 Basis of CI/SfB

An important aspect of the development of CI/SfB has been that the system should be international in outlook – that purely national tables or classification schedules should not be developed or incorporated where satisfactory schedules having international recognition could be used or suitably adapted. The origins of the five tables, each having an international basis, can be shown as follows:

Table 0	Physical environment	Based on UDC
Table 1	Elements	
Table 2	Constructions, forms	Based on SfB Basic Tables of CIB
Table 3	Materials	
Table 4	Activities, requirements	Based on CIB Master Lists

6.2 UDC

UDC is the Universal Decimal Classification system, a general classification particularly suitable for technological subject areas, based on the Dewey system used in public libraries. The UDC schedule below is taken from BS 1000 A:1961 Universal Decimal Classification. Abridged English edition 3rd edition London BSI, 1961. It is reproduced by permission of the British Standards Institution, 2 Park Street, London W1A 2BS.

UDC		CI/SfB	
725.1	Civic and public service buildings	3	Administrative, commercial protective service facilities
725.2	Commercial, office buildings		
725.3	Transport, traffic and storage buildings	1	Utilities, civil engineering facilities
725.4	Industrial buildings	2	Industrial facilities
725.5	Health and welfare buildings	4	Health, welfare facilities
725.6	Prisons, penitentiaries, reformatories		At 3 above
725.7	Public refreshment buildings,	5	Recreational facilities
725.8	Public entertainment buildings		
726	Sacred and funerary buildings	6	Religious facilities
727	Buildings for education, science, art etc	7	Educational, scientific, information facilities
728	Residential buildings	8	Residential facilities

Appendix 6: Correlation with other systems

6.3 SfB Basic Tables of CIB

The SfB system is administered by the Conseil International du Bâtiment (CIB) through its SfB Development Group and its SfB Bureau by means of the SfB Basic Tables. These may be changed only by agreement of the Development Group. The Group consists of representatives of the national licence holders including the RIBA as licence holder for the UK. It also includes a small number of individuals who have special knowledge and experience relevant to the development of SfB for particular purposes. The Basic Tables form the basis for Tables 1, 2 and 3 of this manual and other national application manuals and were last published in 1972 in CIB Report No.22 'The SfB system: authorized building classification system for use in project information and related general information'. They are given in full at the end of this appendix. The main differences between CI/SfB Tables 1, 2 and 3 and the equivalent SfB Basic Tables can be presented as follows:

6.31 Codes which have one interpretation in the Basic Tables and additional interpretations in CI/SfB

Terms in the Basic Tables are to be interpreted only as nouns, eg:

Table 2	Table 3	
Construction	Resource	
F	g2	= clay products (eg: bricks) in brickwork
Brickwork	: Clay (products)	

whereas CI/SfB allows their use either as nouns or as adjectives, eg:

F	g2	= clay brickwork, clay bricks,
Brickwork	: clay	clay bricks or clay products in
Bricks		brickwork or clay brickwork

Using CI/SfB, it must be made clear from the context, or from a special 'role indicator' (eg: L, A, C, see Project information 1.3), which meaning is intended in any particular application of the system. SfB Table 3, as a table of all the resources required for building, includes at a, b, c and d ideas which are at (A), (B), (C) and (D) in Table 4 of CI/SfB. CI/SfB Table 3 should, however, be used in exactly the same way as SfB Basic Table 3 as a classification of resources, if this is right for a particular application of the system.

6.32 Codes which have one definition in the Basic Tables and a different definition in CI/SfB

Basic Tables
(13) Hard floorbeds
(72) Play fittings
G   Prefabricated components for carcass and substructure
X   Prefabricated components for completions, finishes and services

CI/SfB
Hard or soft floorbeds
Play fittings are at (77)
Large block and panel work

Components work

Appendix 6: Correlation with other systems

6.33 Codes used in the Basic Tables which are not used in CI/SfB

Basic Tables
(10) ⎫
(20) ⎬ Site and its subdivision
etc. ⎭

(51) Services Centre

x   Components

CI/SfB
The ideas at (10), (20) etc. in the Basic Tables are given in CI/SfB at (90). For example, (90.2) or (92) in CI/SfB is the equivalent of (20) in Basic Table 1

This idea may be included at (58) in CI/SfB Table 1 where required Code 'x' is vacant in CI/SfB Table 3. It may be used for components if this suits a particular application of the system

6.34 Codes used in CI/SfB which are not used in the Basic Tables

Basic Tables
(18) ⎫ 'Reserved' positions,
(28) ⎬ which may be used
etc. ⎭ as in CI/SfB

(77) 'Free' position which may be used for any purpose

Y   'Reserved' position, which may be used as in CI/SfB

Z   'Reserved' position, which may be used as in CI/SfB

y   'Reserved' position, which may be used as in CI/SfB

CI/SfB
These codes are used for 'other' ideas in each relevant main group, to ensure that there is a place in CI/SfB for every relevant idea

Fittings for special activities (ie implied by ideas in Table 0 such as 'educational fittings')

Formless work and products, products in general

Joints

Composite materials

6.35 Differences in appearance of codes

CI/SfB uses (1–), (2–) etc where the Basic Tables use (1), (2) etc. These Basic Table codes can be used where this suits a particular application of the system.

6.4 CIB Master Lists

Table 4 of CI/SfB is based on the 1972 CIB Master Lists published as CIB Report No.18 'CIB Master Lists for structuring documents relating to buildings, building elements, components, materials and services', just as Table 4 of the 1968 edition was based on the 1964 edition of the Master Lists. The Master List codes are satisfactory for the purposes for which they were designed (to enable cross reference to be made from one Master List to another), but they do not fit into the 'code pattern' of CI/SfB and are somewhat lengthy. For these reasons, and also because certain additions and omissions have been necessary, CI/SfB has continued to use an alphabetical form of code. The presentation below indicates the parallel sequence and the main differences between Table 4 and the Master Lists. The Master List headings are given in more detail at the end of this appendix.

Appendix 6: Correlation with other systems

Master Lists	**CI/SfB Table 4**
0 Document, scope and information for indexing	(A) May be used as direct equivalent of Master List code 0
1 Identification	
1.01 Generic name	
1.02 Product name, type, grade, quality, producer, commodity number	(B) May be used as direct equivalent of Master List codes 1, 1.01, 1.02
1.03 Short description of the product	(C) May be used as direct equivalent of Master List code 1.03
1.04 Related documentation	(D) May be used as direct equivalent of Master List code 1.04
2 Description	Description (E)/(G)
2.01 Constituents	
2.02 Combination of constituents	(E) Composition etc
2.03 Accessories	
2.04 Shape	
2.05 Size	(F) Shape, Size etc
2.06 Weight	
2.07 Appearance etc	(G) Appearance etc
3 Climate, site and occupancy conditions	(H) Context, environment
4 Characteristics related to behaviour in use and working	Performance (J)/(T)
4.01 Structural and mechanical	(J) Mechanics
4.02 Fire	(K) Fire, explosion
4.03 Gases	
4.04 Liquids	
4.05 Solids	(L) Matter
4.06 Biological	
4.07 Thermal	(M) Heat, cold
4.08 Optical	(N) Light, dark
4.09 Acoustic	(P) Sound, quiet
4.10 Electrical	(Q) Electricity, magnetism, radiation
4.11 Energy	
4.12 Side-effects	(R) Energy, other physical factors
4.13 Compatability	
4.14 Durability	
4.15 Working characteristics	see (V) below
4.16 Characteristics relating to maintenance	see (W) below

Appendix 6: Correlation with other systems

5	Application, design	(T)	Application
–	–	(U)	Users, resources
6	Site work	(V)	Working factors
7	Operation and maintenance	(W)	Operation, maintenance factors
–	–	(X)	Change, movement, stability factors
8	Prices and conditions of sale	(Y)	Economic, commercial factors
9	Supply		
10	Technical Services		
11	References	(Z)	May be used as direct equivalent of Master List code 11

6.5 Other SfB Systems

There are two major systems based on the SfB Basic Tables. One of these is CI/SfB, the other is the CBC computer based project management and costing system devised and owned by Bjørn Bindslev, architect and quantity surveyor, Denmark. There are very few code differences between CI/SfB and CBC. Those that there are usually involve codes containing symbols 0, 9, –, A, Z, a, z.

The similarity between the main body of codes in the two systems means that CI/SfB can be used in a similar way to CBC for special applications.

CI/SfB forms the basis for similar systems used in other countries. These include SI/SfB in France, the responsibility of Centre d'Information de la Documentation du Batiment (CIDB) in Paris; BRD/SfB in the Federal Republic of Germany, the responsibility of Stuttgart University. The SfB Agency UK, 66 Portland Place, London W1N 4AD can provide up-to-date information on the situation in particular countries at any particular time.

Appendix 6: Correlation with other systems

6.6 British systems and their relationship to CI/SfB

SFCA The Standard Form of Cost Analysis used by the Building Cost Information Service of the RICS is familiar to all quantity surveyors. Most of the individual element definitions in SFCA correspond with those in CI/SfB Table 1 and the omission of (51) in new CI/SfB, with other changes, brings the two systems still closer together.

BIC The Building Industry Code was developed and promoted by the education consortia, and has thus been used by many local authorities for some part of their building programme. BIC incorporates Tables 0 and 2 of CI/SfB. CI/SfB Table 1 and the list of common element parts in Appendix 1 of this manual are together equivalent to the Elements and Features of BIC.

SMM The Standard Method of Measurement of the RICS consists of work sections based either on construction form concepts like 'Brickwork' as found in CI/SfB Table 2; or on elemental concepts like 'Drainage' as found in CI/SfB Table 1. It has been found that bills of quantities in which the content of each page is the lowest common denominator of Tables 1 and 2 of CI/SfB can be re-sorted into SMM work sections without further division.

CIT The Construction Industry Thesaurus was developed in 1970 by the Polytechnic of the South Bank under the sponsorship of CIRIA and DOE, as an aid to post-coordinate indexing and as a means of controlling and improving terminology used throughout the construction industry. Consistent arrangement of items and documents is very important if benefit is to be gained from using CI/SfB. Consistent terminology is less important. In the case of CIT this order of importance is reversed. CIT has been of great value in improving both classification and terminology in this edition of CI/SfB, and has also led to the development of Appendix 1. The main categories of CIT and CI/SfB correspond broadly as follows:

CI/SfB	CIT
Physical environment	Construction works
Elements, Constructions, Forms	Parts of construction works
Materials	Materials
Activities	Operations
Requirements	Properties, processes
Peripheral subjects	Peripheral subjects

SfB Basic Table 1: Elements

Appendix 1 to CIB Report No. 22 'The SfB System'

Site
- (00) Reserved
- (10) Prepared site
- (20) Site structures
- (30) Site enclosures
- (40) Roads, paths, pavings
- (50) Site services (piped and ducted)
- (60) Site services (electrical)
- (70) Site fittings
- (80) Landscape, play areas
- (90) Summary site
  Information applicable to two or more of the groups (10) to (80). All general information applicable to the site as a whole.
- (–0) Reserved

(1) Ground substructure
Parts below underside of lowest screed or where no screed exists to underside of lowest floor finish.
- (10) See site
- (11) Ground, earth shapes
- (12) Free
- (13) Floor beds
- (14) Free
- (15) Free
- (16) Foundations (other than pile foundations)
- (17) Pile foundations
- (18) Reserved
- (19) Summary ground substructure (building)
  Information applicable to two or more of the groups (11) to (18).
- (1–) Reserved

(2) Structure
(including carcass)
- (20) See site
- (21) Walls, external walls
- (22) Internal walls, partitions
- (23) Floors, galleries
- (24) Stairs, ramps
- (25) Free
- (26) Free
- (27) Roofs
- (28) Reserved
- (29) Summary structure (building)
  Information applicable to two or more of the groups (21) to (28).
- (2–) Reserved

(3) Completions
(to structure including openings in structure)
- (30) See site
- (31) External walls' completions
- (32) Internal walls' completions
- (33) Floors', galleries' completions
- (34) Stairs', ramps' completions
- (35) Suspended ceilings
- (36) Free
- (37) Roofs' completions
- (38) Reserved
- (39) Summary completions (building)
  Information applicable to two or more of the groups (31) to (38).
- (3–) Reserved

(4) Finishes
(to structure)
- (40) See site
- (41) Wall finishes externally
- (42) Wall finishes internally
- (43) Floor finishes
- (44) Stair, ramp finishes
- (45) Ceiling finishes
- (46) Free
- (47) Roof finishes
- (48) Reserved
- (49) Summary finishes (building)
  Information applicable to two or more of the groups (41) to (48).
- (4–) Reserved

(5) Services
(mainly piped, ducted)
- (50) See site
- (51) Services centre (mainly piped, ducted)
- (52) Drainage, refuse disposal
- (53) Liquids supply services
- (54) Gases supply services
- (55) Space cooling services
- (56) Space heating services
- (57) Ventilation and air conditioning services
- (58) Reserved
- (59) Summary services (mainly piped and ducted, building)
  Information applicable to two or more of the groups (51) to (58).
- (5–) Reserved

(6) Services
(mainly electrical)
- (60) See site
- (61) Electrical centre
- (62) Power distribution services
- (63) Lighting services
- (64) Communication services
- (65) Free
- (66) Transport services
- (67) Free
- (68) Reserved
- (69) Summary services (mainly electrical, building)
  Information applicable to two or more of the groups (61) to (68).
- (6–) Reserved

(7) Fittings
- (70) See site
- (71) Display, circulation fittings
- (72) Rest, work, play fittings
- (73) Culinary, eating, drinking fittings
- (74) Sanitary, hygiene fittings
- (75) Cleaning, maintenance fittings
- (76) Storage, screening fittings
- (77) Free
- (78) Reserved
- (79) Summary fittings (building)
  Information applicable to two or more of the groups (71) to (78).
- (7–) Reserved

Reserved groups other than those previously mentioned
(01)/(09) and (0–)
(81)/(89) and (8–)
(91)/(99) and (9–)
(–1)/(–9) and (––)

SfB Basic Table 2: Constructions

A	Preliminaries, general conditions, general costs and general work including general temporary work	P	Thick coating work
		Q	Free
B	Demolition and shoring work	R	Rigid sheet work
C	Excavations and loose fill work	S	Rigid tile work
D	Reserved	T	Flexible sheet work (except L)*
E	Cast in-situ work	U	Free
F	Block work	V	Film coating and impregnation work
G	Works of prefabricated components for carcass and substructure	W	Planting work (work with living forms)
H	Section work	X	Works of prefabricated components for completions, finishes and services
I	Pipe work		
J	Wire work, mesh work	Y	Reserved
K	Quilt work*, insulation work	Z	Reserved
L	Flexible sheet work (proofing)*	—	Reserved
M	Malleable sheet work*		
N	Rigid sheet overlap work*		
O	Reserved		

* Including work with tiles, i.e. small sheets

SfB Basic Table 3: Materials and other resources

a **Administration**
(or general)

b **Plant, tools**

c **Labour**

d Reserved

e/o Formed materials

e **Natural stone**
- 0 General
- 1 Granite, basalt, other igneous
- 2 Marble
- 3 Limestone (other than marble)
- 4 Sandstone
- 5 Slate
- 6 Reserved
- 7 Reserved
- 8 Reserved
- 9 Other formed natural stone materials

f **Pre-cast with binder**
- 0 General
- 1 Sandlime concrete (precast)
- 2 All-in aggregate concrete (precast), heavy concrete (precast)
- 3 Terrazzo (precast)
- 4 Lightweight, cellular concrete (precast)
- 5 Lightweight aggregate concrete (precast)
- 6 Asbestos-based materials (preformed)
- 7 Gypsum (preformed)
- 8 Magnesian materials (preformed)
- 9 Other materials pre-cast with binder

g **Clay (dried, fired)**
- 0 General
- 1 Dried clay e.g. sun dried, unburnt
- 2 Fired clay, unglazed fired clay
- 3 Glazed fired clay, vitrified clay
- 4 Reserved
- 5 Reserved
- 6 Refractory materials, heat resistant materials
- 7 Reserved
- 8 Reserved
- 9 Other dried or fired clays

h **Metal**
- 0 General
- 1 Cast iron, wrought iron
- 2 Steel, mild steel
- 3 Steel alloys, e.g. stainless steel
- 4 Aluminium, aluminium alloys
- 5 Copper
- 6 Copper alloys, e.g. brass, bronze
- 7 Zinc
- 8 Lead
- 9 Other metals, e.g. tin, chromium, nickel
Other metal alloys

i **Wood** (including impregnated wood)
- 0 General
- 1 Timber (strength graded, unwrot)
- 2 Softwood (in general, and wrot)
- 3 Hardwood
- 4 Wood laminates, e.g. plywood laminboard, blockboard, faced wood laminates
- 5 Wood veneers
- 6 Reserved
- 7 Reserved
- 8 Reserved
- 9 Other wood materials (other than those at j1, j7, j8)

j **Vegetable and animal materials**
- 0 General
- 1 Wood fibres, e.g. fibre building board, hardboard
- 2 Paper, corrugated paper
- 3 Vegetable fibres (other than wood)
- 4 Reserved
- 5 Bark, cork
- 6 Animal fibres, leather
- 7 Wood particles (chipboard)
- 8 Wood wool-cement
- 9 Other formed vegetable and animal materials

k Reserved

l Reserved

m **Inorganic fibres**
(in preformed products)
- 0 General
- 1 Mineral wool/fibres (e.g. rock wool), glass wool/fibres, ceramic wool/fibres
- 2 Asbestos wool/fibres
- 3 Reserved
- 4 Reserved
- 5 Reserved
- 6 Reserved
- 7 Reserved
- 8 Reserved
- 9 Other formed inorganic fibrous materials

n **Rubber, plastics, etc**
- 0 General
- 1 Asphalt (preformed)
- 2 Impregnated fibre and felt
- 3 Reserved
- 4 Linoleum
- 5 Rubbers (natural and synthetic)
- 6 Plastics, synthetic fibres
- 7 Cellular plastics
- 8 Reserved
- 9 Other rubber, plastics materials. Mixed natural/synthetic fibres

o **Glass**
Sub-division free

SfB Basic Table 3: Materials and other resources

p/s Formless materials

p **Aggregates, loose fills**
0 General
1 Natural fills, aggregates
2 Artificial aggregates in general
   Artificial granular (heavy)
3 Artificial granular (light)
4 Artificial ash
5 Shavings
6 Powder
7 Fibres
8 Reserved
9 Other aggregates, loose fills

q **Lime and cement binders, mortars, concretes**
0 General
1 Lime
2 Cement
3 Lime-cement (mixed hydraulic binders)
4 Mortars, concretes in general lime-cement-aggregate mixes
5 Terrazzo, granolithic mixes
6 Lightweight, cellular concrete mixes
7 Lightweight aggregate concrete mixes
8 Reserved
9 Other lime-cement-aggregate mixes, asbestos cement

r **Clay, gypsum, magnesia and plastics binders, mortars**
0 General
1 Clay mortar mixes
2 Gypsum, gypsum mixes
3 Magnesia, magnesia mixes
4 Plastics binders and mortar mixes
5 Reserved
6 Reserved
7 Reserved
8 Reserved
9 Other binders and mortar mixes (except those in q)

s **Bituminous materials**
0 General
1 Bitumen, pitch, tar, asphalt
2 Reserved
3 Reserved
4 Mastic asphalt
5 Stone/bitumen mixes (aggregates bonded together with bitumen)
   Rolled asphalt, coated macadam
6 Reserved
7 Reserved
8 Reserved
9 Other bituminous materials

t/v Functional materials

t **Fixing and jointing agents**
0 General
1 Welding materials
2 Soldering materials
3 Adhesives, bonding materials
4 Joint fillers, putty, mastics
5 Reserved
6 Fasteners
7 Ironmongery
8 Reserved
9 Other fixing and jointing agents

u **Protective, process/property modifying agents**
0 General
1 Anti-corrosive materials
2 Modifying agents, admixtures
3 Rot proofers, fungicides, germicides, insecticides
4 Flame retardants
5 Polishes, seals, hardeners, size
6 Water repellents
7 Reserved
8 Reserved
9 Other protective, process/property modifying agents (other than v), e.g. anti-static agents

v **Paints**
Sub-division free, should include varnishes, lacquers, paint fillers, pigments etc.

w **Ancillary materials**
0 General
1 Rust removing agents
2 Reserved
3 Fuels (gases, liquids and solids)
4 Water
5 Acids, alkalis
6 Fertilisers
7 Cleaning materials, abrasives
8 Reserved
9 Other ancillary materials

x **Components**
(by function)

y Reserved

z Reserved

— Reserved

CIB Master list

From CIB Report No. 18, pages 51–2

0 Document, scope and information for indexing
.01 Scope of document
.02 Classification
.03 Key words
.04 Organization responsible for information given
.05 Date of publication and period of validity of document

1 Identification
.01 Generic name
.02 Product name, type, grade, quality, producer, commodity number
.03 Short description of the product, its purpose and conditions of use, together with any limitations of use
.04 Related documentation
   Statutory (legal) requirements
   Standards
   Quality and assessment certificates
   Guarantees
   Codes of practice; national specifications

2 Description
.01 Constituents, parts, type of finish
.02 Combination of constituents and parts
.03 Accessories
.04 Shape
.05 Size
.06 Weight
.07 Appearance, including texture, colour, pattern, capacity, lustre; feel; smell

3 Climate, site and occupancy conditions

4 Characteristics relating to behaviour in use and working
.01 Structural and mechanical; strength and deformations
.02 Fire
.03 Gases
.04 Liquids
.05 Solids
.06 Biological
.07 Thermal
.08 Optical
.09 Acoustic
.10 Electrical
.11 Energy
.12 Side-effects
.13 Compatibility
.14 Durability
.15 Working characteristics
.16 Characteristics relating to maintenance

5 Applications, design
.01 Suitability; functional
.02 Suitability; economic
.03 Suitability; statutory
.04 Resource conservation
.05 Design details
.06 Design specification clauses
.07 Mistakes in use

6 Sitework
.01 Labour, plant, materials and space requirements
.02 Work planning
.03 Work off site
.04 Transport, handling and storage
.05 Preparatory site work
.06 Work on site; erection, site assembly, finishing
.07 Protective measures
.08 Cleaning up
.09 Site quality control
.10 Labour safety and welfare
.11 General public safety and welfare

7 Operation and maintenance
.01 Labour, plant, materials and space requirements
.02 Method of operation and control
.03 Cleaning and maintenance
.04 Repair and replacement
.05 Labour safety and welfare
.06 General public safety and welfare

8 Prices and conditions of sale
.01 Purchase price
.02 Contract conditions
.03 Terms of payment

9 Supply
.01 Sources of supply and supply capacity
.02 Packaging, labelling
.03 Directions for ordering
.04 Conditions of delivery

10 Technical services
.01 Servicing and maintenance organization and facilities
.02 Technical advisory services

11 References
.01 Location of samples in use
.02 Literature